WATCHING BIRDS

WATCHING BIRDS

by JAMES FISHER

Revised edition by

JIM FLEGG

With drawings by Crispin Fisher

Great Spotted Woodpecker

T. & A. D. POYSER

Berkhamsted

First published 1974 by T & AD Poyser Ltd.
Print-on-demand and digital editions published 2010 by T & AD Poyser,
an imprint of A&C Black Publishers Ltd, 36 Soho Square, London W1D
3QY

www.acblack.com

Copyright © 1974 by T & AD Poyser

ISBN (print) 978-1-4081-3866-3
ISBN (epub) 978-1-4081-3867-0
ISBN (e-pdf) 978-1-4081-3865-6

A CIP catalogue record for this book is available from the British Library

This is a print-on-demand edition produced from an original copy.

It is produced using paper that is made from wood grown in managed
sustainable forests. It is natural, renewable and recyclable. The logging and
manufacturing processes conform to the environmental regulations of the
country of origin.

Printed and bound in Great Britain

Contents

List of plates

From the Preface to the 1940 edition

Some people might consider an apology necessary for the appearance of a book about birds at a time when Britain is fighting for its own and many other lives. I make no such apology. Birds are part of the heritage we are fighting for. After this war ordinary people are going to have a better time than they have had; they are going to get about more; they will have time to rest from their tremendous tasks; many will get the opportunity, hitherto sought in vain, of watching wild creatures and making discoveries about them. It is for these men and women, and not for the privileged few to whom ornithology has been an indulgence, that I have written this little book.

I shall be very pleased if anybody who reads it becomes interested in birds.

Oundle, November 1940   JAMES FISHER

Preface to 1974 edition

A great many people over three decades have been introduced by *Watching Birds* to the tremendous fields of interest that opened out before them once their attention has been drawn to the possibilities. Beyond the pure, largely aesthetic, pleasure of just looking at birds are to be found the delights of *watching* birds and attempting to understand more of their lives, their movements or their biology. As an interest, birdwatching is compulsive: amongst its charms are that as a hobby it can be practised anytime and anywhere, and that the commonest of birds retain their charm and fascination no matter what level of expertise is reached.

This much was realised by James Fisher when he first devised this book. He could hardly have visualised the impact that it would have: the fantastic growth of birdwatching and bird conservation societies is in large part due to his influence, and for many years the mainstream of fieldworkers involved in the studies of the British Trust for Ornithology was initially stimulated by his abilities to present general ideas in an attractive manner.

The structure he planned remains quite adequate, though the attendant techniques may have progressed farther than even James Fisher could visualise. Where my own ideas differ from his, I have tried to state in an open discussion our points of view, and in many cases I have retained outdated figures because of the interesting comparisons they offer between the 1930's and the 1970's.

It is a daunting task to revise the bird book on which you cut your teeth: it is the surest measure of the man who wrote it that what is needed, after thirty-odd years, is an updating and not a sweeping revision.

JIM FLEGG

Tring, January 1974

Note on the illustrations

Some illustrations in the book are based on those in the original edition, although for the most part the original illustrations have been revised and redrawn, but James Fisher's thanks to the various workers responsible for the basic data must be reiterated. For some of the distribution maps John Parslow's book, *Breeding Birds of Britain and Ireland*, has been an invaluable source of information in bringing details up to date.

It would have gratified James Fisher greatly to see how many of the new figures in this edition are based on the results of the cooperative fieldwork of members of the British Trust for Ornithology published in *Bird Study, B.T.O. News* and other journals. The value of such products of combined enthusiasm is perhaps best seen in the B.T.O. interim *Atlas* maps, the maps of ringing recoveries and in the bird population Indexes. To all the fieldworkers involved, and to the B.T.O., I express my gratitude.

Crispin Fisher has put his artistic talents, with excellent effect, to the bird drawings in this edition, and by common consent we have retained some line drawings from the original edition so that father and son may appear together. The majority of the photographs, including that on the cover, are my own but my thanks are due to Richard T. Mills for so ably remedying the deficiencies in two widely separated fields.

J.J.M.F.

1 Introducing the birdwatcher to the bird

"My remarks are, I trust, true in the whole, though I do not pretend to say that they are perfectly void of mistake, or that a more nice observer might not make many additions, since subjects of this kind are inexhaustible."

GILBERT WHITE, *9 December 1773*

THE BIRDWATCHER

Though they are by no means the most ubiquitous, numerous, curious, or diversified forms of life, it is nevertheless a fact that birds have had more attention paid to them than any other corresponding group of animals. Many attempts have been made to discover why so many people like birds, but nobody has so far given a really satisfactory answer. When it comes, it will probably be found partly from a sort of mass observation and partly from the long and detailed history which birdwatching has enjoyed, particularly in Britain. Certainly birds are relatively obvious and always about us, but they are extremely mobile in contrast with other beautiful and interesting natural things like flowers.

All sorts of different people seem to watch birds: James Fisher knew of a Prime Minister, a Secretary of State, a charwoman, two policemen, two kings, one ex-king, five Communists, four Labour, one Liberal, and three Conservative Members of Parliament, the chairman of a County Council, several farm-labourers, a rich man who earns two or three times their weekly wage in every hour of the day, at least forty-six schoolmasters, and an engine-driver. To these can be added today (in reflection of both political and cultural shifts of emphasis) two Princes Consort, at least two Presidents, numerous peers, a Field Marshal and many people falling under the general heading of 'show business personalities'. The number of birdwatchers is now impossible to estimate—our only guide to its enormity is the membership of the various formal clubs and societies, some local, some national, which must now be approaching the half-million mark. Obviously, these are only a small fraction of those with a genuine interest in, and feeling for, birds.

The reasons which make these men and women interested in birds are, surely, very divergent. Some, when asked, can give no particular reason for their liking for birds. Others are quite definite—they like their shape, their colours, their songs, the places where they live; their view is the aesthetic one. Many of these paint birds or write prose or poetry about them. Still others, like several of the schoolmasters, are grimly scientific about them, and will talk for hours on the territory theory, the classification of the swallows, or changes in the bird population of British woodland during historic times. The luckiest ones are those who blend the two: retaining the aesthetic pleasure and adding to it the delights realised by an inquisitive interest.

The observation of birds may be a superstition, a tradition, an art, a science, a pleasure, a hobby, or a bore; this depends entirely on the nature of the observer. Those who read this book may give bird watching any one of these definitions; which is a sound reason why we should get down as soon as possible to the business of introducing the bird-watchers to the birds.

THE BIRD

Birds are animals which branched off from reptiles at roughly the same time in evolutionary history as did the mammals. Though both birds and mammals have warm blood, from anatomical considerations it is clear that these two branches are quite separate, and that the birds of to-day are more closely related to reptiles than they are to mammals. The parts of the world that birds live in are rather limited. Though birds have been seen in practically every area of the world from Pole to Pole and at heights more than that of Mount Everest, yet they have rarely been found to penetrate more than about 30 feet into the earth or about 150 feet below the surface of the sea; and every year some part of their lives has to be spent in contact with land, for birds cannot build nests at sea, even though some freshwater species build floating nests and rarely step onto dry land.

In the hundred and thirty million years or so in which birds have existed as a separate class of animals, they have retained their basic structure to a remarkable extent, and, compared with the situation in other classes of animals, even the apparently most widely distinct birds are internally little different. To avoid going into anatomical details, it must be enough to say that the same bones (that is, homologous ones) exist in the wings of humming-birds and ostriches, and in the feet of hawks and ducks. What has happened in the course of evolution is that these have been modified for different purposes. The wing of the humming-bird has become perhaps the most efficient organ of flight in the animal kingdom; (the humming-bird is, as far as we know, one of the

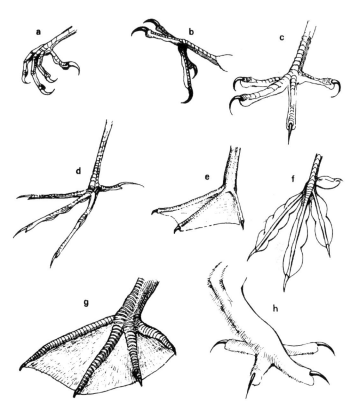

1   Birds' feet adapted to differing modes of life: (a) finch family (passerine birds)—perching; (b) woodpecker—tree climbing; (c) Sparrowhawk—seizing prey in flight; (d) Water Rail—load-spreading; (e) Puffin—(three-toed) swimming; (f) Coot—(lobed) swimming and load-spreading; (g) Shag—(four-toed) swimming; (h) Barn Owl—gripping prey

few birds which can actually fly backwards); while that of the ostrich, which can no longer fly, is used as an organ of balance and of display. The talons of the hawk, provided as they are with terrible claws, grasp its prey, and are usually the weapons which cause the victim's death (rather than the beak); while the feet of the duck are webbed so that it can walk upon mud and swim. Figure 1 gives just some idea of the immense variety of foot adaptation and Figure 2 shows some of the adaptations of beaks to feeding.

Many are the ways in which birds are adapted to their surroundings,

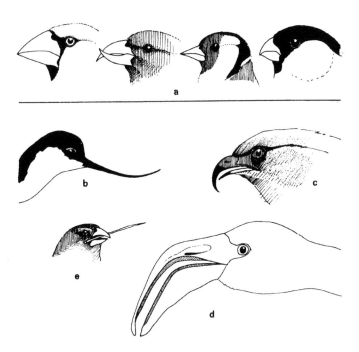

2   Birds' beaks adapted to special feeding methods: (a) Hawfinch, Crossbill, Goldfinch and Bullfinch—seed crushing and extraction; (b) Avocet—skimming for small molluscs, crustacea, etc; (c) Everglade Kite—(snail eating) shell piercing and extraction; (d) flamingo—filtering plankton, etc; (e) Galapagos Woodpecker Finch—tool holding (stick to "winkle" insects from holes)

and to the lives which nature (or, if you like, evolutionary history) has taught them to live. Many interesting adaptations can be seen in British birds: so that they may climb trees and cling to them while hunting insects, the woodpeckers have feet with two toes pointing forwards and two aft, and the under-side of their tail feathers is rough where these rest against the trunk of the tree as the birds climb; their beaks are like chisels, for often they carve much of their nest out of the living wood, and their food habits take the great beaks under bark and into crevices in their search for insects. The central tail feathers of woodpeckers are specially strengthened to serve as a prop while the bird feeds, but those of the Nuthatch, a bird of very similar habits, are not. The Nuthatch, however, unlike the woodpeckers, can climb both head-up and head-down because it is not dependant on support from its tail. In North America one group of woodpeckers has gone even farther, not so much in adaptation as in the habits produced by such an adaptation. They have deserted insect catching for sucking the sap of the trees—hence their name of sap-suckers—and this is a warning to the unwary: birds are not always what they seem, and adaptations may hide relationships, or indicate false ones.

Many birds are beautifully adapted to their surroundings at the earliest stage of their lives: the egg. The eggs of the Ringed Plover and of other shore birds often resemble very closely the pebbles among which they are laid (Plate 14); those of birds which build open nests are often coloured with a protective pattern of pigment. But it is a biological economy to lay an egg without pigment where pigment is unnecessary. The eggs of woodpeckers and owls are white when they are laid, and coloured later on only by dirt, and these birds nest in holes where their eggs cannot be seen by enemies. Oddly enough, the egg of the Puffin still has grey or red spots, like faint copies of those of its cousin, the Guillemot, which rather suggests that the Puffin has only recently (in the evolutionary sense) taken to nesting in burrows. One of the nicest adaptations of all is shown by the Guillemot, an auk which nests on very narrow cliff ledges. The egg of the Guillemot is conical—almost pear-shaped—and if knocked, rolls in a tight circle on its narrow end, thus staying on the ledge when an egg of normal shape would surely have rolled off.

When we speak of adaptation of the sort which we have just described, we must remember that our attention is likely to be drawn to the more remarkable examples of adaptation to special needs and away from the fact that practically every organ and part of an animal is adapted (directly or indirectly, efficiently or inefficiently, in greater or lesser degree) to the animal's surroundings, or to its demand at some period of its total or daily life. A bird is what it is partly because of the immediate environment—since food or weather may affect its colour or size—partly because of heredity—which provides machinery, producing in turn variations on which natural selection can act—partly because of chance—since chance appearance of such new variations as are ignored by natural selection may determine some of the characters of an individual or even of an isolated group of individuals. We must remember that not only special kinds of beaks but beaks in general are adaptations. A wing is an adaptation for flying. An egg is an adaptation for nourishing the young. Adaptations, in birds as in other animals, must be regarded in terms of the perspective of the ages, the dynamics of evolution, and the complicated mechanism of heredity. We may regard any characteristic as adaptive if it can be explained as fitting the bird to its mode of life.

HOW A BIRD IS BUILT

The skeleton of a bird (Fig. 3) is, in general plan, the same as that of other vertebrate animals. Essentially, it consists of a strong box made up of many bony elements, with a variety of attachments. The head, first, is attached to a column of strong articulating vertebrae, the backbone. To the axis provided by the head and vertebrae are attached various appendages; to the head the jaws; to the lower vertebrae of the neck from one to four "floating" ribs; to the vertebrae of the back, ribs which curve forwards to join the huge breast-bone, forming the strong case inside which the lungs and heart lie, and to the outside of which is attached, towards the back, the shoulder girdle; to the remaining vertebrae the hip girdle. The shoulder girdle supports the elements of the wings, the hip girdle those of the legs.

The number of vertebrae in birds is most variable, unlike that in mammals. A swan may have twenty-five bones in its neck, a sparrow only sixteen. (It is odd to recollect that a giraffe, in common with man, has only seven.) In the region of the back some pigeons may have only three vertebrae whose ribs meet the breast-bone, while ducks, swans, gulls, rails or auks may have seven or eight. In the last group the thorax (the box formed by ribs and breast-bone) is very long, and the un-

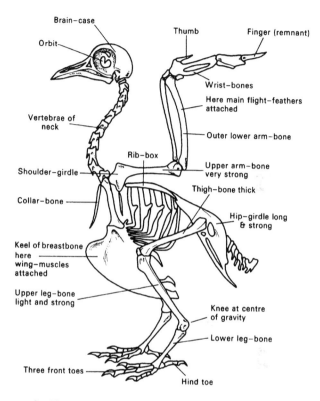

Brain-case

Orbit

Thumb

Finger (remnant)

Wrist-bones

Here main flight-feathers attached

Vertebrae of neck

Outer lower arm-bone

Rib-box

Shoulder-girdle

Upper arm-bone very strong

Thigh-bone thick

Collar-bone

Hip-girdle long & strong

Keel of breastbone here wing-muscles attached

Upper leg-bone light and strong

Knee at centre of gravity

Lower leg-bone

Three front toes

Hind toe

3   Bird skeleton with wings raised—common names given

protected abdominal region between it and the base of the tail is very short—a useful adaptation towards a swimming or deep-diving life.

One of the chief characteristics of a bird's bones that we should expect to find is that of lightness, an adaptation towards flying. Though birds may have large bones, these have not a massive structure. Most are hollow—some have huge cavities which may contain extensions of the air-sacs of the lung—and only occasionally do we find cross-struts (as in the albatrosses). These bones owe their strength and rigidity to a structure so sophisticated as to be emulated by engineers in modern girder design. They are almost unbelievably lighter than a mammal bone of the same size.

Though there are many characteristics of a bird's skull which show its evolutionary affinities rather than the tasks to which it is adapted, yet it is better for us, in this short review, to consider it from an adaptive point of view. There are several requirements which a bird's skull must fulfil. First, it must be light; secondly, the loss of teeth and manipulating fore-limbs during the course of evolution must be compensated; thirdly, the brain and eyes must be accommodated and protected. The first point is covered by the fact that few skull-bones of birds are more than plates and struts; the second by the high development of the horny beak, which has to act almost as a limb, and is, in practice, a very effective one; the third by the huge orbits, which leave only a very thin partition between the eyes on each side. These restrict the rest of the brain to the back of the skull—which is accordingly broad. That part of the brain which deals with smell (of which birds have very little sense) is much reduced; that which deals with co-ordination of movement and balance is understandably large; that which deals with what are sometimes called the "higher functions" (such as logic, or aesthetic appreciation, in humans) is small. Nonetheless, some of these "birdbrains" (so wrongly called) are capable of navigating in most weathers, with really pinpoint accuracy, to a wintering place in Africa, and in the case of the Swallow, back here to the very same barn or garage to nest in the succeeding summer.

The greatest differences between the skeletons of birds and those of other vertebrates are found in the skull and in the pectoral girdle and forelimb. The latter corresponds to the shoulder-blade, collar-bone, arm, forearm, and hand of man. The pectoral girdle is attached to the bones of the main trunk at only one point, where the front end of the shoulder-blade is attached, on each side, firmly to the fore-part of the breast-bone. The two shoulder-blades are braced together across the front by the collar-bones, which are fused in the middle to form what we call the wish-bone. Behind, the free ends of the shoulder-blades are bound by stiff and strong ligaments to the ribs and to the vertebrae of the back.

When the wing is working the strain comes directly on to the box formed by vertebrae, ribs, and breast-bone which are together called the thorax. This box has therefore to be extremely strong. It is. At the back the vertebrae are fused together, at the sides the ribs (often with

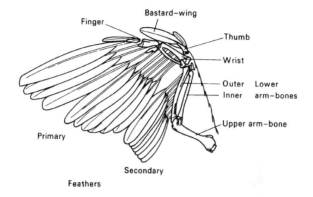

4   Skeletal wing from below

strengthening bony processes) are bound to each other by further ligaments, and in front the breast-bone is built on engineering principles that combine strength with great lightness plus a keel for the attachment of the powerful muscles of the wing.

This is the basis upon which the wing can operate. The part that flaps (Fig. 4) is composed of three units. Attached to the shoulder-blade by a ball-and-socket joint is the upper arm-bone (humerus), a single strong rod. To the other end of the humerus are joined the two bones (radius and ulna) of the forearm, and to them in turn is joined the third unit, several small bones, largely fused together (carpals and metacarpals), representing the mammalian wrist and hand.

The muscular force which works the wing is derived almost entirely from huge wedges of muscle attached from the wish-bone in front to the base of the breast-bone below and the shoulder-blade behind. The fibres of these great muscles converge to an insertion on the humerus. By far the larger part of these pectoral flight muscles is concerned with the downbeat of the wing: inside these, connected to the humerus by a complex but subtle tendon system, are the smaller muscles that lift the wing for the next flap. Next time you carve the breast of a chicken, or better a pheasant, look out for these two muscle groups.

It is the humerus, then, which takes the strain of the wing beat, though the feathers of the wing (Fig. 5) are carried not on this but on the other bones of the arm. The secondary feathers are attached to the ulna,

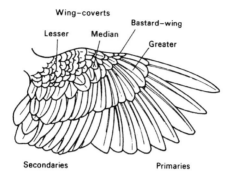

Wing-coverts

Lesser Median Bastard-wing Greater

Secondaries Primaries

5 Wing feathers from above

the lower and thicker of the two bones of the forearm; these feathers play the major part in propulsion. The primary feathers are attached to an elongated wrist, and to two bones which correspond to the index finger; these play an effective part in steering. Thumb and third finger are present in rudimentary form, the former bearing a bunch of feathers often called the bastard wing, which appear to have the aeronautical function of a wing slot. The other fingers are absent.

All other attempts at flight by vertebrates (and there have been upwards of half a dozen) have involved the development of flaps of skin, stretched, as in the pterodactyl, from the little finger to the feet, or, as in the bat, over all the outstretched fingers to the feet (Fig. 6). In no case except birds have feet been kept out of the mechanism; besides which, skin is clumsy to fold compared with feathers. So it can be seen that birds have a considerable advantage over other flying vertebrates. The extent of this advantage can be gauged from the fact that pterodactyls are extinct and bats restricted in range and habits, whereas birds are everywhere.

Because birds have been able to keep their feet out of their flying mechanism, these organs have been left free for the play of evolutionary adaptation. But because birds can fly, there are some considerable problems to be got over before they can run. The great pectoral muscles that work the wings may weigh, in the pigeon, over a quarter of the weight of the whole body. The centre of gravity of a flying bird lies well in front of the joint between leg and body. Hence, if the bird is to stand

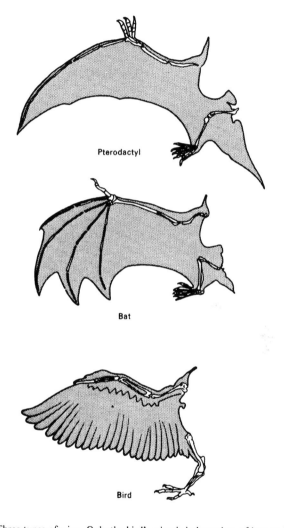

6   Three types of wing. Only the bird's wing is independent of legs and tail. Note that the pterodactyl uses one finger only for the wing frame, but the bat employs four, with the thumb for clinging

upright, the girdle of the hip must clasp the vertebrae in a vice-like grip. This, of course, is what it does. In front of and behind the socket into which the head of the thigh-bone inserts, the central part of the hip-girdle is elongated and fused with the vertebrae. It will be remembered that the vertebrae of the back were fused together the length of the thorax, to give a strong base for the working of the wing. The last one or two of these rib-bearing vertebrae are also integrated with the hip girdle, and are clasped by the broad iliac bone. Below them, perhaps ten or eleven more vertebrae are fused with the girdle, and beyond these two or three more little ones may project; these are the last of the caudal or tail vertebrae, and represent what there is of a true tail in modern birds. So the total result is that the bird has a very rigid back from the base of its neck to its tail, a back whose rigidity is achieved by the fusion of vertebrae to each other, or to the hip girdle, forming one of the main elements of this strong, box-like structure which not only protects the vital organs but also withstands the stresses and strains of both flight and walking.

What appears to be the thigh of a bird is in reality its shin, and what looks like its knee the wrong way round is its ankle. The true thigh-bone (femur) is usually short, and runs almost horizontally forwards to the knee, which may lie closely applied to the surface of the body just about under the centre of gravity, tucked well away into the body feathers and out of sight.

Running downwards and slightly backwards, the true shin (apparent thigh) is a single long bone (tibio-fibula); all that is left of the second leg-bone found in most land vertebrates is a sliver applied to the surface of the main bone. The next member runs downwards and forwards, a single bone (tarso-metatarsus) derived from elements originally part of the ankle and the upper ends of three of the toes. The toes themselves are never more than four in number; the first (big toe) is nearly always turned backwards when it is present. The fifth (little toe) is entirely absent.

It can be seen that the bird's leg, composed as it is of three single rigid bones, jointed together, provided with toes at the end, is ideal for its purpose, and can be adapted for running, standing, taking landing strain, and catching prey. The knee- and ankle-joints of the bird lie some distance from the hip-joint and toe-joints, and work in opposite directions. Thus they constitute perhaps one of the most effective shock-absorbing mechanisms in the animal world.

THE SOFT PARTS

Anybody who has prepared a fowl for the table will know, and those who have carved one will suspect, that the "guts" of a bird (Fig. 7) are contained in a cavity stretching from rib box to hip girdle, and covered very largely by the downward projection of the breast-bone. This cavity is generally known as the abdominal cavity, and in it the long continuous tube of the alimentary tract is coiled. Food is taken in by the beak, laced with saliva from glands in the mouth, and squeezed down the throat by muscles at the back of the mouth. In the throat region the tract is called the gullet or oesophagus. It may be expansible and used to store food, or it may have a large special bag attached to it for storage; if present, the latter is known as the crop. Crops may sometimes be used as "udders"; those of pigeons secrete a milk-like protein-rich fluid used to feed the young. Sometimes they may be used as extra gizzards: that of the Hoatzin has muscular walls which can squeeze the juices out of its food.

The gullet continues through the thorax, passing behind the heart, to

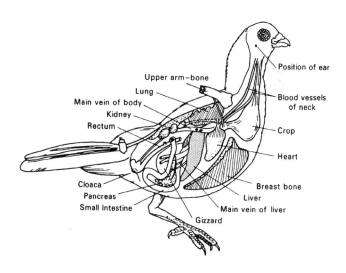

7   The internal organs (pigeon)

reach the abdominal cavity, where it immediately broadens out into a large bag, the stomach. Usually, the first part of this has walls supplied with glands which secrete acid in order to continue the processes of digestion started by the saliva, and the after-part is lined with muscle and developed as the gizzard. Many birds which live on grain or on hard vegetables have large gizzards and frequently swallow small stones, which are retained in the gizzard as an aid to crushing. Others which live on flesh have no gizzards at all, but have glands over the whole of the stomach; and in some birds of prey, for example, the gizzard serves as a collecting area for bones and fur which are later "cast" up as pellets.

After the stomach, the tract continues as a loop known as the duodenum. Into this run three ducts from the pancreas, and down the ducts run secretions which change starch to sugar and break down proteins and fats to simpler forms. Ducts also run into the duodenum from the liver, which may be even larger than the stomach, which it often surrounds. The top surface of the liver is closely applied to the diaphragm, the broad muscular wall which divides the cavity of the thorax (containing the heart and lungs) from that of the abdomen. The liver serves as storage for excess sugars, to secrete bile to aid digestion, and as a general "cleaning up" organ where toxic materials are removed from the blood.

After the duodenum we reach the small intestine proper. This is often very long, especially in vegetable-feeding birds. In it digestion continues, and the main part of the absorption of the digested products takes place. It loops and coils about the abdominal cavity, each loop being held in place by a membrane attached to the back wall, and practically fills the cavity. Finally, it joins the rectum, a wide short tube which leads to the cloaca. At the junction of the small intestine with the rectum we find one or two blind tubes (caeca) branching sideways. Sometimes these are quite large, and it is possible that matter such as cellulose is broken down in them by bacterial action and rendered fit for absorption. It is thought, too, that water and proteins may also be absorbed.

The final passage, the short cloaca, communicates with the exterior, but before this it receives the excretory ducts from the kidneys and the reproductive ducts from the testes or ovaries. Thus the cloaca is a general-purpose opening; it is both excretory and reproductive, and is the only such opening in the lower part of the bird's body.

THE HEART AND BLOOD

Though the heart of birds is built in a different way from that of mammals (it has more affinities with the reptile heart), it is four-chambered, and apparently is just as efficient, since in it oxygenated blood is effectively separated from the de-oxygenated blood which has delivered up its oxygen to the body. Birds have a more rapid heart-beat than mammals, and their temperature is considerably higher—from about 37°C. in birds like the emu and ostrich to about 44°C. in some small larks and finches. Like mammals, the temperature of their body is more or less constant and independent of the temperature of their surroundings, rising and falling only slightly with periods of activity or rest. As in mammals (e.g. bats), so there are, in some birds, some exceptions to this rule. Under certain conditions of the environment, such as the cold nights of the high Andes, the maintenance of body temperature by small birds like humming-birds (which have a relatively large surface in proportion to their weight) becomes difficult or impossible. Some humming-birds, under these circumstances, become torpid. At first their breathing becomes violent, and they lose the power of flight. When touched, they make a peculiar whistling sound. When completely torpid they are quite rigid, the head is pointing upwards and the eyes are closed, the breathing appears to stop. It may take the birds about half an hour to recover from this stage. Recently, the quite astonishing discovery was made in America that the Poor Will (a nightjar-like bird) can hibernate for long periods in rock crevices to survive the winter. Under these conditions of torpidity, the body temperature drops to about 5°C. and the general rate of bodily processes falls to about three per cent of normal.

Thus under certain exceptional circumstances birds can revert to the reptile-like condition of having a body temperature that is not constant. In adult birds this condition is very rare, is only found in certain species, and, when present, is an adaptation to special circumstances of the environment. On the other hand, the young (in the nest) of most birds show a variable body temperature. This is because the mechanism of temperature control does not get into full working order until most of the feathers have grown sufficiently to produce an efficient insulating layer stopping a continual heat loss. Hence the care which many birds take, by brooding and, equally important, by shading from the sun, in order to protect their young from excessive cold or heat.

The rapid heart-beat of the bird is a necessary adaptation, not only towards keeping up its temperature, but towards supplying the huge demands of the flying muscles. These require sugars, and oxygen to "burn" them with, for the large amounts of energy needed. And the blood has to carry the necessary supplies from the liver—the sugar storehouse—and the lungs. The heart itself is large, and takes up more space in the thorax than does the heart of a mammal; its walls are very thick and muscular, for it has a lot of work to do in pumping the blood round the body—blood which carries, besides sugar and oxygen, the waste gas carbon dioxide (the product of the "burning" of the sugars), other waste products, proteins for body building, fats for storage and more energy, the secretions of the ductless glands (chemical compounds often called hormones which act as "messengers"), and various bodies concerned with the destruction and neutralisation of bacteria and poisons.

THE LUNGS

A bird's windpipe opens at the back of the tongue and runs down the throat to the top of the thorax. Here it broadens out into the voice-box or syrinx. This consists of a main chamber, with a bony band inside. This band has two processes attached to it on which a membrane is stretched; there may be other membranes present as well, and together these act as vocal cords, producing sound and song. Not surprisingly, the whole structure in noted songsters is much more complex than simple quacking mechanism of the ducks. The nearly voiceless Ostrich has no real syrinx at all, nor have the vultures. From the bottom of the voice-box two tubes branch off, one to each lung. These tubes feed not only the lungs but, through them, nine "air sacs" which are often large. These sacs increase the amount of air available inside the bird for use—very useful if the species is a great songster or a great diver. By means of a complex routing of the air (not yet properly understood) these air sacs also fill the lungs directly after an out-breathing, so that air is available for respiration in the periods before it can be breathed in from outside, giving birds the ability to extract oxygen whilst breathing out and in—an exceedingly efficient technique and one that maintains the vital energy needs of the flight muscles. And since birds cannot sweat, a lot of the water exchange which helps to keep their temperature under control takes place over the inner surface of these sacs.

THE REPRODUCTIVE ORGANS

Placed within the abdominal cavity, closely applied to the back wall on each side of the mid-line, lie a pair of flattish, lobed organs. These are the kidneys, and from the middle of each a slender tube runs direct to the cloaca. The kidneys filter waste products from the blood and excrete them in semi-liquid solution through the ducts.

Most female animals have a pair of ovaries slung from the abdominal wall close to each kidney, but most birds have (for some reason) only the left ovary; if the right is present, it is nearly always a vestige. The ovary looks like a bunch of grapes; its size varies greatly, from enlarged with forming eggs in the breeding season, to near-invisible in winter. A long, thick, coiled tube which runs to the cloaca on the left side has at its upper end a great funnel which opens directly into the abdominal cavity near the ovary. This is the oviduct (Fig. 8).

When the breeding season arrives, a chain of events is set in motion inside the bird, usually by changes in temperature and daylength. Certain ductless glands begin to pour their secretions, often under the influence of the pituitary—a special ductless gland in the region of the brain—which appears to act in some ways as a master gland. Changes

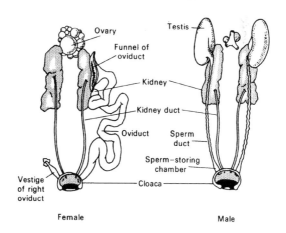

8   The reproductive organs of female and male birds

begin to take place in the ovary (which is a ductless gland in its own right, producing hormones like the others), and the unfertilised and immature eggs begin to take in yolk and get bigger.

In the male bird organs are attached to the back wall of the abdominal cavity near the top of the kidneys in much the same place as the ovary in the female. In the male, however, they are paired and of equal size and function. These are the male reproductive organs; oval bodies, tiny in the off-season but huge at breeding time. They consist of a part known as the testis, in which the sperms and some hormones are produced, and another part mainly concerned with storage of the male fluid. From each organ runs a narrow duct, of about the same size as the tube from the kidney, to the cloaca.

When birds copulate, the female generally takes up a crouching position and lifts her tail. The male mounts on her back and applies his cloaca to hers. By muscular action, a dose of male seminal fluid is then pumped into the female's cloaca from the sperm-ducts, where it has been stored. Some male birds which might otherwise have difficulty in securing efficient transfer of the sperm (such as ducks, which often copulate on or under the water), have a special muscular penis attached to their cloaca which fits into that of the female. Not many birds, however, are provided with such an organ.

DEVELOPMENT

The sperms of the male birds, of which there are many thousands in a dose of fluid, are living things with an independent existence, and have many of the properties of separate organisms. One of these properties, that of swimming against a current, now comes into play; and the sperms make their way, against the slow movement of the secretions from the walls of the oviduct, up this tube. They may meet an egg or eggs either at the top of the oviduct or in the oviduct itself; when they do, they fuse with the egg. As soon as a sperm has entered an egg, the egg's surface changes chemically and no other sperms can get in. The egg, once fertilised, rapidly begins to divide. It already has a supply of yolk and consists of a dividing, developing group of cells above, and a mass of nutritious yolk below.

The next step is for the yolk with the developing embryo upon it to be wrapped in a safe and useful parcel. This is done by cells in the walls of the top of the oviduct, and the first wrapping consists of an albuminous substance which, in the final full-sized egg, can be seen as a sort of curly part of the white between the yolk and the ends of the egg. When the egg is laid and the shell is hard, this curly part acts as a kind of spring shock-absorber suspending the yolk, hammock-like, safely in the middle of the albumen, or "white". As the egg passes farther down the oviduct, more of the white of egg is secreted and deposited round it. Still farther down the tube are cells which secrete the chalky shell, and farther on still are more cells whose job it is to produce pigments to make the typical pattern on the egg. The egg is finally laid, broad end foremost, by contractions of muscles in the wall of the cloaca.

The embryos of animals more simple than the hen—those of frogs, for instance—develop by division of the fertilised egg and rapidly form a little round mass which then becomes something like a yolk-filled cup. The embryos of birds (Fig. 9) cannot develop in this way because of the great size of the yolk. So instead of growing as a roughly ball-shaped mass, they start life as a plate of cells draped over the top of the yolk. To begin with, this plate is streak-shaped, and very soon after the egg has been laid the first cells of the nervous system can be seen in the mid-line. After forty hours in the chicken, the new animal is over a quarter of an inch long, is more sausage-shaped than streak-shaped, and is beginning to separate a little from the yolk. This separation is most pronounced in the front third of the growing body. This part is going to be the head, and already swellings can be seen where the eyes are to be; down the rest of the body little round dots indicate the segments of the growing bird. Four days from laying we can distinguish brain, eye, ear, and heart, the last as a sort of bag surprisingly near the head. Plates of muscle are already arranging themselves in the segments of the body. After a week the eyes have become really large and the limbs have begun to appear. All the same, it would be difficult to tell whether an embryo was that of a chicken, a crocodile or a cat.

By now the chick is getting big and its growth demands a supply of oxygen. This has to be carried in blood-vessels, and large ones develop in the membranes which surround the chick. During the next day or two little cones appear all over the surface of the embryo; these are the buds of developing feathers. By the tenth day we can tell that we have a young bird, because the limbs are of the proper bird shape and structure, and we can see the beginnings of a beak. At the end of a fortnight quite a lot of real bone has been laid down in the skeleton in place of soft cartilage, and the muscles are well developed. The beak has been

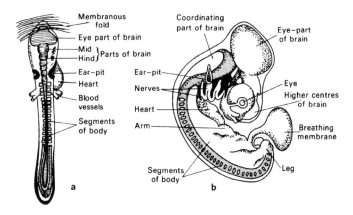

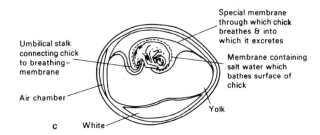

9   Three stages in the development of the chick: (a) 36 hours after the egg is laid (back view); (b) five days after laying (side view), limbs beginning to appear, breathing membrane developed and eyes well advanced; (c) nine days after laying, embryo well developed

provided with an egg tooth with which the chicken can break its way out when the moment of hatching comes about a week later; the feathers have begun rapid growth and are much longer, and the head is less huge in proportion to the rest of the body.

The period of time between laying and hatching is known as the incubation period. It may be as long as forty or fifty days in certain seabirds, ostriches and brush-turkeys. In the hen it is three weeks, in some small perching birds only ten or eleven days.

When the time for hatching comes, the chick pushes its beak into the air chamber at the broad end of the egg and takes a first fill of air into its lungs. Up to now its oxygen supply had come from the air diffusing through the walls of the shell and received by the network of blood-vessels, outside the chick itself. Having filled its lungs with air, the chick then strikes the egg shell and breaks it. After an hour or two it may have broken off the broad end of the egg and may have struggled into the outside world. Helping it to chip away the shell is the small, hard protuberance on the end of the beak, the "egg tooth", which can still be seen in many young birds for two or three days after hatching.

In general there are two kinds of young birds. Nidifugous (literally, "fleeing the nest") chicks can run as soon as they are hatched and hatch at a comparatively advanced stage. Game birds, wading birds, and ducks have nidifugous chicks, all very active and capable of running, swimming, and hiding. Their eyes are open and functional, and often they can at least partly feed themselves. The most striking example of a nidifugous bird is the young of the brush-turkey, a Megapode from Australia and nearby countries. The egg of this bird is incubated, not by either parent, but by the heat of fermentation of a pile of decaying vegetable matter scratched up by the male bird. The young often have to find their own way out of a pile of stuff, and as soon as they have managed to do this, they can flutter off the ground on wings which have reached a remarkably high stage of development. On the other hand, birds which nest in holes or which build complicated and comfortable structures can afford to have their young hatched at an earlier period of development; these are, then, nidicolous. A newly hatched tit or blackbird has scarcely any developed feathers at all and is naked, ugly (almost reptilian), to all intents and purposes quite blind, and helpless.

It would be wrong to end this introduction to birds without brief reference to their feathers, which are, after all, one of their most distinctive features. No other animal possesses feathers. In its early embryological stages, the feather closely resembles the reptilian scale from which it has evolved. There are a variety of feather types: most birds have their body insulated by an "underclothing" layer of filamentous down, covered by the contour feathers that we see, which themselves have fluffy, insulating bases. The wing and tail feathers have much the same basic structure as the contour feathers, but are specially strongly developed.

Feathers are made of the protein keratin, and are secreted by special

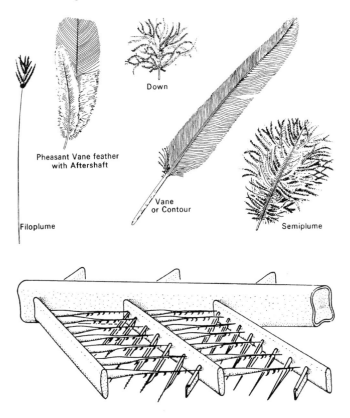

Pheasant Vane feather
with Aftershaft

Down

Filoplume

Vane
or Contour

Semiplume

10   Various types of feathers and a detail of a vane feather to show the hooked barbules which interlock the feather barbs

cells in the skin. Figure 10 shows how extraordinarily complex their structure is when looked at closely, and how important preening must be to the bird, for the mechanical properties of the feathers (and thus weatherproofing and flight) depend on the interlocking of the tiny hooks on the barbules. Naturally structures of this sort are subject to wear and tear, and most small birds change all their feathers, in a process called moult, at least once and sometimes twice each year.

2 Arranging the birds

"Faunists, as you observe, are too apt to acquiesce in bare descriptions, and a few synonyms: the reason is plain; because all that may be done at home in a man's study, but the investigation of the life and conversation of animals is a concern of much more trouble and difficulty, and is not to be obtained but by the active and inquisitive, and by those that reside much in the country."

GILBERT WHITE, *1 August 1771*

SPECIES

There are about eight thousand five hundred species of birds in the world, according to the authorities. This is really quite a small number; there are over a hundred thousand kinds of molluscs and about three quarters of a million kinds of insects (of which a third are beetles), and entomologists are still describing thousands of new species of insects every year. Yet there are probably more ornithologists than there are malacologists or entomologists in the world.

It is likely that birdwatchers are able to satisfy their keenness so well mainly because of the interest taken in birds by taxonomic zoologists. Taxonomic zoologists usually inhabit museums and spend most of their lives examining, arranging and sorting skins and skulls. These painstaking people have one of the most difficult tasks in the whole of zoology, for in their arrangement of their animals in a natural classification they have to bear in mind many very different principles. When a collection of bird skins turns up from some part of the world that has not been very well worked, the taxonomist has to be an historian, an evolutionist, a geographer, an anatomist, a bibliographer, and often a mathematician before he can be justified in assigning his specimens to their proper species or if necessary to a new species. Even then, if he has all these qualities, he may often have to rely on an intuitive flair for diagnosis.

In the Linnaean system (a system of classification published by Linnaeus in 1758) every animal bears two names, a generic one and a specific one. The same generic name is not permitted by the international rules of nomenclature to occur twice in the animal kingdom, though specific names may often be repeated. Once a name has been given to an animal by an authority, with a description in a recognised scientific journal or some other such printed paper, it sticks unless somebody can later prove that a name has already been given to the animal concerned or that the animal does not merit being classed as a new species. This rule usually works well and saves a great many mistakes and inconveniences. Sometimes, however, it has awkward results. For instance, the names of the British Song Thrush have been changed several times in the last twenty years because of new discoveries about the dates of publication of the original papers in which it was described. The Manx Shearwater is called *Puffinus puffinus* and the Puffin *Fratercula arctica*. This seems curious to British birdwatchers, but has to be, since the name *Puffinus* was originally given to the shearwater.

It should be realised that most modern taxonomists have two main points of view when deciding what a species is. They have to look at their animal, perhaps first from the point of view of practical convenience, giving each animal a label, or "pigeonhole", so that it may be easily and accurately referred to again; and secondly they have to look at it from the evolutionary aspect. What they like to have as a species is a collection of animals which their colleagues would readily agree to be such, and which, when met with alive in the field, would have some reality to the animal watcher. It is often extraordinarily hard to decide whether a population of animals, including birds, really is all of one kind, either because of the natural variability in biological material, or because of a phenomenon called convergent evolution (when quite different species may develop similar features to suit life in a special environment).

A simple definition of a "species" is: two animals belong to the same species if they inhabit the same area, have the same tradition of structure, colour, voice, and habits, and tend to breed with each other rather than with similar animals which do not have those traditions.

Two animals do not necessarily belong to the same species if they interbreed in the wild. There are many examples of distinct species which have increased their range in the course of evolution so as to overlap. In the region of overlap they may interbreed, producing a mixed or hybrid population. Nevertheless, this does not mean that they are of the same species.

Two animals belong to different species if, when inhabiting the same area and having much the same outward form, they avoid interbreeding by devices of colour, voice, or display, or by selecting different habitats in that area. As we will see in a later chapter, Chiffchaffs and Willow Warblers are very similar and are quite hard to distinguish even in the hand. In nature they remain distinct species because of their distinctive songs (by which they recognise each other) and probably by slight differences in their choice of habitat.

A great range of size, colour, and geographical distribution does not destroy the reality of a species. Thus, if you compared a Puffin from Spitsbergen with one from southern Europe, you might easily conclude that they were of different species until you had seen a collection of birds intermediate in size from the intervening coast of the north Atlantic. As a matter of fact there is a beautiful gradation in size, increasing from south to north. Taxonomists have given names to the populations in various regions up the coast, and these populations can all be arranged in order of size and latitude. Not one of these populations could be called a different species, but they can be quite suitably called different subspecies or geographical races. It would be possible to go even farther and say merely that there was a gradation of "cline" in size from south to north.

Sometimes a population of animals of one species may originally have been spread over a large area. In the course of time and for various reasons (perhaps an ice age) this population may have been broken up into isolated groups. As time goes on, such groups, breeding among themselves, may begin to show differences from other groups, which become more and more marked as time incı ƶases. Such is the birth of a subspecies. As soon as the average difference between one group and the next becomes consistently and obviously greater than the average difference between one individual and another within a group, we are justified in calling each of these groups a subspecies. When these differences become really great, sufficiently so for interbreeding to become impossible for one reason or another, we are forced to go farther and call the groups different species.

Many examples of such separation are found on islands, and in Britain there are quite a lot of island forms of birds. Thus there is a mainland Wren, a Shetland Wren, a Hebridean Wren, and a St Kilda Wren (Fig. 11). When the St Kilda Wren was first described, it was given the rank of a full species, but zoologists have now decided that it should only be called a subspecies. Perhaps in a thousand years or so it may become even more different from the mainland form and may be

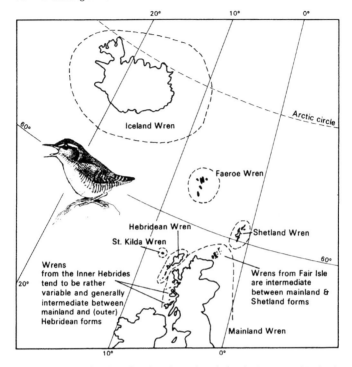

Iceland Wren

Faeroe Wren

Shetland Wren

Hebridean Wren

St. Kilda Wren

Wrens
from the Inner Hebrides
tend to be rather
variable and generally
intermediate between
mainland and (outer)
Hebridean forms

Wrens from Fair Isle
are intermediate
between mainland &
Shetland forms

Mainland Wren

Arctic circle

11 Wren—distribution of various forms in Britain, the Faroes and Iceland

worth calling a species again. The borderline between species and sub-species is (in the case of island forms) extremely difficult to draw. The St Kilda Wren, for example, has wings about 10% longer than its mainland counterpart, is considerably heavier, and very grey, only the tail being the rich rufous brown of the mainland bird. The song, too, is held by some to be different, but this is difficult to judge because the towering St Kilda cliffs make a very different sounding board to an English woodland!

Some subspecies of birds are not purely geographical ones since there are other barriers which can isolate populations of birds. In America one ornithologist who happened to be a very keen field worker dis-covered that two races of Clapper Rails overlapped in broad geographical range. He was unable to explain why they had distinctive characteristics of form until he found that one race lives on marshes and the other in dry country. Here, then, is an example of two races separated ecologically, that is, by their natural ways of life.

It is probably not worthwhile for the ordinary amateur field ornithologist, or birdwatcher, to trouble very much about the definition of the species he is working with. In Britain, at all events, the work has already been done very intensively, and although there is still a bit of a battle going on between those who like to recognise similarities between races of birds ("lumpers") and those who like to recognise differences ("splitters"), this need not concern us very much. As far as British birds are concerned, the amateur can rest assured that the list of species is in the hands of a very capable committee of the British Ornithologists' Union. The committee meets regularly to discuss very carefully all the additions and subtractions necessary for the British list, and all the changes of name that bibliographical and anatomical research have made necessary. If the reader wants to know the exact names of the species on the British list he should consult the B.O.U. book *The status of birds in Britain and Ireland* (Oxford, 1971). The additions and corrections are published annually in the *Ibis*, which is the journal of the British Ornithologists' Union, and if the birdwatcher uses these lists and with their aid examines his experience in the field, he will get a better idea of the reality of species and subspecies than any series of random, theoretical definitions could give him. It is important nevertheless for him to realise that improvements in our knowledge of classification can come from his own work. Birds are notoriously hard to classify properly, and for help in their task zoologists are having to depend on every new source of ideas that they can touch. Today, the systematist has to consider not merely things like measurements, structure, and colour, but also voice, habits, geographical position, migration, population, type of nest, egg size and shape, and now even the chemical structure of the egg-white proteins. In many of these aspects, the amateur birdwatcher can help greatly.

VARIATION

No two animals are alike. It is impossible to say that two animals are truly the same from the zoological point of view, except in the rare case of identical twins (and this is not really a safe case). We have already traced some general trends of mass variation, such as the trend in the Puffin towards bigness as it gets north; and there are many more examples. Thus, birds from moist regions tend to be darker than those from dry regions, and in cold parts of the world exposed parts like feet and beaks tend to be reduced in size and often covered with feathers. But this kind of organised and correlated variation is very different from individual variation. Sometimes certain individuals of a species are very different from the norm; for instance, they may lack pigment altogether, in which case they are called albinos, or they may have an excess of, or altogether lack, some but not all of the normal pigments. These variants are called mutants and are rare things, and the more extreme kinds probably play very little part in the evolution of a species, although in some moths black ('melanistic') mutants play a novel role in the modern

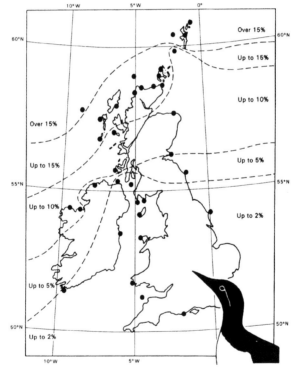

12   Guillemot—distribution of the bridled form as a percentage of total breeding stock around the British Isles

world in camouflaging the individual more successfully on grimy, sooty tree trunks.

Sometimes, however, variants appear very regularly, so that there is an apparent (and often a real) balance of numbers between variant and norm. Such a variant is present in our Guillemot, for certain individuals have a white "bridle" or "spectacle" circling and running backwards from the eye. Among the breeding stocks of Guillemots in Britain, the bridled birds (Fig. 12) increase from under a quarter per cent of the total population in the south of England, up to 26 per cent in Shetland. Beyond this northern point of Britain the percentage increases still further until it reaches 53 per cent in the south of Iceland and probably more on Bear Island. Clearly there is a definite gradation in space of the proportion between bridled and normal Guillemots. This gradation, or cline, is not perfectly continuous from north to south; there are a few inconsistencies round the coasts of Britain, and bridled birds are less common in the north than in the south of Iceland. All the same, it is broadly true to say that the number of bridled Guillemots increases as one goes north.

The Fulmar is a bird in which there are several colour phases; it may show every gradation from almost white on head and underparts to deep, smokey grey all over, in which plumage they are called "blue". The farther one goes from Britain towards the east coast of Greenland, the darker become the Fulmars. Again, this colour-phase-cline is not a simple south to north affair; it is possible that the birds actually get lighter northwards up the west coast of Greenland, and the situation in the European arctic seas of Spitsbergen is obscure. A lot of information about the distribution of the colour phases of the Fulmar has been collected from ships at sea, and late in 1939 the Admiralty were kind enough to issue a Fleet Order asking officers to collect information on this subject and send it to the British Trust for Ornithology, resulting in the distribution map in Figure 13.

We must also remember that in birds (as, for example, in man) there is variation in size between individuals. Sometimes this is associated with the sex of the individual. In most birds, the males are larger and longer-winged than the females (the birds of prey are an exception to this) but, over an above this, while most individuals of a species are reasonably close to the mean, exceptions do occur, in rapidly diminishing numbers as the deviation from the mean becomes more extreme. Figure 14 shows how this occurs in the Water Rail.

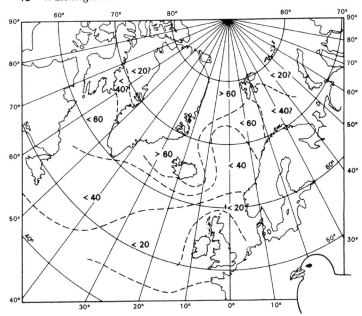

13   Fulmar—distribution of "dark" phase Fulmars in the North Atlantic in summer as a percentage of total breeding stock

THE HIGHER CLASSIFICATION

Further up the taxonomic scale, above species and subspecies, stand genera, containing a number of related species. Related genera are gathered into *Families* and related families are gathered into *Orders*.

Let us give an example. The British Twite is recognised as a geographical race, resident in the British Isles. It has been named *Carduelis flavirostris pipilans* (Latham). Latham was the authority who in 1787 gave the name *pipilans* to this particular bird. According to the International Rules of Nomenclature, his name has to be put in brackets because he originally attached the name *pipilans* to a different genus (*Fringilla*), from which the bird has now been removed.

The British Twite belongs to the subspecies *pipilans* of the species *flavirostris* (to which belongs also the subspecies *Carduelis flavirostris flavirostris*, the Continental Twite, and others) of the genus *Carduelis* (to which belongs also the species *Carduelis cannabina*, the Linnet, and others) of the Family *Fringillidae* (to which belong also the genera *Coccothraustes* (Hawfinch), *Pyrrhula* (Bullfinch), and others) of the Order *Passeriformes*, or perching birds (to which belong also the families *Corvidae* (crows), *Sturnidae* (starlings) and others).

From the ordinary fieldworker's point of view the important thing is the bird. He is more likely to be interested in the species than in the genus. The one thing essential to him is the power to identify correctly. From whatever point of view he may approach birdwatching, he needs skill in identification. This skill is as vital to him if he has the aesthetic approach as if he has a scientific one. If he cannot identify he will find it

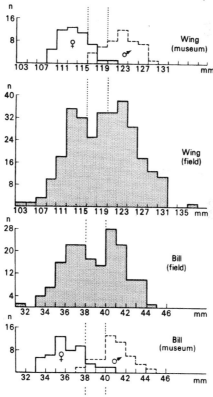

14   Wing and bill lengths of Water Rail, male and female, in the field and as museum specimens

difficult to enjoy himself with birds. If he begins to watch birds straight off with the help of one of the more old-fashioned text-books, he may soon be in trouble, because he will be asked to look at birds from the standpoint of the museum taxonomist, and the museum man has his own special way of describing birds. He can describe every feather but not every pose, and can describe the structure of the voice-box but say nothing about the voice. The reader may be given detailed descriptions of all that is permanent in a skin, but less good descriptions of things like the colour of the eye and legs, and poorish descriptions of any voice or habits.

However, with today's wide choice the problem is one of making a selection from the tremendous field of identification guides available. There have been great strides in recent years in presenting bird illustrations and descriptions compactly, drawing attention to vital identification features. It is worth remembering that these giant strides forward have largely been based on the fieldwork of amateur enthusiasts, who have faced the problems of putting a name to the bird in the field, analysed them, and gone a long way to solving the other problem of transferring some of their knowledge and skills to others.

Let us trace roughly the course of the average birdwatcher as he discovers the delights and defeats of identification.

1. He becomes interested in birds, perhaps because of the influence of somebody else.

2. He cannot make them out, especially when he is alone.

3. So he gets a book with pictures, which may be of two types, (a) rather accurate pictures, or (b) pictures designed not as pictures but as aids to identification, particularly in regard to colour.

4. Armed with these he improves his knowledge of birds.

5. He begins to get acquainted, not merely with the birds, but also with their habits.

6. With the further weapon of habits he begins to realise the *nuances* of identification and can speculate about them.

7. Realising this, he probably widens his circle of ornithological friends.

8. He then becomes capable of identifying birds, not merely by the text-book colours, but also by flight, song, and all sorts of minor habits of the most subtle kind, called by most birdwatchers "jizz".

9. He is then in a position to contribute to our knowledge of bird identification.

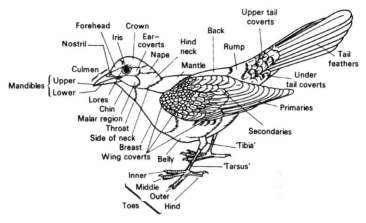

15 The topography of a bird

Such is the evolution of the birdwatcher's attitude towards identification. It is not by any means our purpose, in a short, introductory book such as this, to try to give a key to the British birds, when there are so many books serving this purpose already on the market. What must be said, however, is that you will not get anywhere watching birds without practice, and that you should try to strike a balance between field observations of your own and the notes you take, and the use of text-books.

If you are puzzled by a new bird you see, it is not always the best plan to rush straight off to the text-book. You may have one in your pocket, and the best thing to do is to write accurate notes on every characteristic you can, for comparison with the text-book description later. Watch the new bird for as long as possible, note the sort of places it likes, what food it is seeking, its call note or song (though this is difficult, often, to transcribe), its type of flight, and of course its colour, shape, and so on. Try to estimate its size in comparison with any other birds that are about. Any colours or adornments which are clearly of use to the bird for signalling purposes, like white marks on the wings, white rumps or red breasts, are probably equally useful to you in identification. It is very useful to have a fair knowledge of the technical terms for the various parts of the bird and the various tracts and rows of feathers. Figure 15 names the parts in general use, and though some of the names may seem to be a little cumbersome, they are really worth remembering.

3 The tools of birdwatching

"In the last week of the last month five of those most rare birds, too un-common to have obtained an English name, but known to naturalists by the terms of *himanotopus* or *loripes*, and *charadrius himanoptus*, were shot upon the verge of Frensham-pond, a large lake belonging to the bishop of Winchester, and lying between Wolmer-forest and the town of Farnham in the county of Surrey. The pond keeper says there were three brace in the flock; but that, after he had satisfied his curiosity, he suffered the sixth to remain unmolested."

GILBERT WHITE, *7 May 1779*

There are a number of general tools that can be considered as essentials to the birdwatcher—binoculars, a notebook, and some guides to bird identification, for example. There are others, like suitable clothing, a camera, a small library of bird books, which while not vital can considerably increase his comfort, skill and overall enjoyment of the hobby. Beyond this, the more involved he becomes in some fields of bird study, the more sophisticated and specialist is the equipment needed: rings and traps for marking, a tape recorder for bird song, 25″ to the mile maps for census work and so on. Lying beyond even this are cars and inflatable boats at one end of the scale and nestboxes and dummy eggs in a variety of colours at the other.

OPTICAL EQUIPMENT

This seems the sensible place to start, because any form of birdwatching can only be improved by the choice and use of suitable binoculars which will bring the action nearer or allow detailed study of colour and plumage. However, it should never be forgotten that in some circumstances—geese against an estuary sky or a "smoke column" of waders shimmering over the mud—binoculars can *restrict* the view and detract from the overall beauty of the scene. But, in general, with binoculars you not only stand a reasonable chance of recognising your bird, but you can see, and wonder at, what it is doing. We are immensely fortunate that modern technology has made available, at reasonable price, a wide (perhaps bewilderingly so) choice of binoculars. What points should you bear in mind when making your choice? There are some reasonably simple rules:

1. QUALITY: generally speaking, you get what you pay for—your dealer should advise you on this. Remember that an expensive buy may last a lifetime, but you do not have the opportunity to change your mind (unless you are very rich!) or to obtain two pairs of glasses for different situations.

2. MAGNIFICATION: most birdwatchers find ×6 too low—×7 ×8 ×9 are good general-purpose magnifications. For watching large reservoirs, estuaries or the sea, × 10 or × 12 will be more satisfactory, but may well be too heavy to be held steady by a feminine hand. Variable magnification ("zoom") binoculars are usually heavy and often suffer from poorer definition.

3. FIELD OF VIEW: should be as large as possible, and tends to be better with lower magnifications. A wide-field is most helpful in woodland or similar situations. Try out rival brands in the field.

4. LIGHT GATHERING: most binoculars have both magnification and object glass (the larger lens) diameter marked on the body—e.g. 8 × 30, 10 × 50. A quick and simple way of estimating the light gathering capabilities of binoculars (that is, how good they are likely to be in the poor light of early morning or evening) is to divide the object glass diameter by the magnification. So for 8 × 30, this becomes 3.75 and for 10 × 50 it is 5.0. The higher the figure, the better. To obtain similar evening performance to the 10 × 50, 8 × 40 would be needed.

5. OPTICAL PERFORMANCE: this *must* always be checked in the field before finally purchasing. Check that you can hold and focus the glasses easily; that there is no distortion of vertical lines (look at a pole or pylon); that there are no colour fringes ("rainbow" halos round the image). Some slight lack of focus round the edges of the field is normally acceptable, but the central area (where you will usually concentrate) must be crystal-sharp.

Now, what are the right binoculars for you? If you are birding mostly in woodland or heathland areas, a lower magnification, wide-field and good light gathering will be needed. For the wide-open spaces like an estuary, then × 10 is more likely to be your choice. If a compromise is essential then perhaps × 9 fits the bill, but do consider the possibilities of

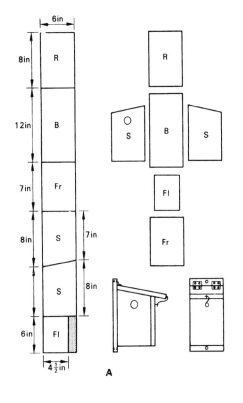

A

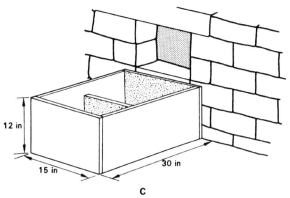

C

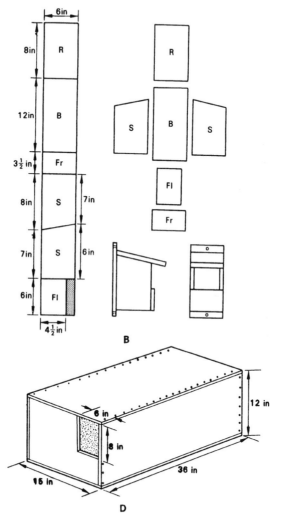

16 Plans for nestboxes, from the BTO Field Guide No. 3, *Nestboxes*—(a) hole nesters, small (tits, Nuthatch, Pied Flycatcher, etc); (b) open-fronted, small (Robin, Spotted Flycatcher, Pied Wagtail, etc); (c) large tray (Barn Owl, Kestrel, feral pigeon, etc); (d) Barn Owl box—with the supply of natural nest sites for the Barn Owl declining you can help conserve the species with appropriately placed nestboxes of this type

becoming the proud owner of *two* pairs of binoculars, each suited to your precise needs.

For even higher-powered work, like the close examination of "difficult" waders or the identification of distant ducks, a telescope can be used—although this instrument takes a little getting used to! The light transmission is poor, allowing the use of a telescope only in fairly good light conditions. The field of view is tiny, and spotting your bird takes practice. Nonetheless, modern telescopes represent a considerable advance on those featured so often in birdwatching cartoons. Focusing is now by a twist-grip, rather than by pulling and pushing on brass draw-tubes, and a similar twist-grip allows the magnification to be changed (usually over a range ×20–×60 or similar) without loss of focus. So you find your bird on lowest power, and then "zoom" up to get the best available view.

CAMERAS

A camera is by no means the necessity that binoculars are, but it is often useful and can be, on occasions, a highly rewarding accessory. Photographs—especially colour transparencies—of places you have visited and birds you have seen not only provide a valuable record, but also untold enjoyment when viewed again, sometimes years later. Here too, modern technology has brought really sophisticated equipment within the reach of most of us, and it is by no means necessary for the beginner to limit himself solely to habitat shots. The price and availability of equipment lead me to suggest that although a larger format gives higher quality, there is no real alternative to a 35 mm single-lens-reflex (SLR) camera. With these, the picture you see in your viewfinder is the one that appears on the film, so that you can be sure that your subject is both correctly positioned and in focus. Many SLR cameras now have the exposure metering system incorporated in the viewfinder, which helps considerably with wildlife photography. Most SLR's have easily interchangeable lenses (the Pentax/Praktica screw is perhaps the commonest) so that you can quickly attach a long-focus (telephoto) lens for bird close-ups if you so desire. A word of warning though—a 200 mm lens may suffice if you are working from a hide on largish birds, but for "stalking" even garden birds without a hide, at least a 300 mm lens is needed.

CLOTHING

This is another area where personal comfort and preferences hold considerable sway, and where sometimes a compromise must be struck. Despite this, there are some essentials and some limits that may be given.

If you are birdwatching in remote, or exposed areas, especially if (as in Britain) the weather may suddenly deteriorate, and if you are going to do any hill-walking or rock scrambling, then you *must* be properly equipped. In these circumstances a twisted ankle could be as serious as two broken legs, so take no risks. Always have adequate warm, weatherproof clothing and some food and drink. Always have properly robust and properly soled footwear for the job. Always carry at least one brightly coloured item of clothing, and in really remote areas, a whistle.

Otherwise, the basic rule is to be warm (or cool) enough, and reasonably weatherproof. Totally waterproof clothing or footwear usually results in excessive condensation, so you finish up wet anyway! Other than in the circumstances I have already outlined, the birdwatcher should always be inconspicuous (and this applies to his *behaviour* as well as his clothes). Large numbers of more or less waterproof, insulated anoraks are available for winter wear, and thick trousers keep your legs warm. Nylon over-trousers keep out cold winds as well as rain. A good compromise for the spring and autumn (long periods often embracing summer also in Britain), and one that I favour myself for its quick-change opportunities and easy portability, is a combination of thick roll-neck sweaters and a nylon "kagoul" (a sort of hooded smock). This allows you to keep your body temperature where you want it, and to be totally wind or waterproof when necessary, all without carrying a huge rucksack for your wardrobe.

If at all possible, avoid clothing that rustles or crackles as you move, or which scrapes noisily past twigs and leaves. Much of the art of getting close views of birds lies in effective stalking; quietness, steady and unhurried movements, judgement of light and wind, and a steadily built-up knowledge of the meanings of the other noises of the countryside and the likely behaviour of the bird that you wish to see, all play their part.

It is astonishing just how much many people miss as they walk through the countryside. Chattering noisily, scuffling through leaves and snapping twigs, walking across the middle of open ground or along the top of a bank—all are bound to give the birds the maximum warning of your arrival, and to restrict your views to distant ones. Always try and use the available cover to mask your progress—keep behind hedges or on the fringe of clearings, keep below the top of the river bank or sea-wall, pausing cautiously from time to time to peer over the top—it is worth it for the better view. If you can, keep the sun behind you or over your shoulder, as colours change or vanish if you must look into the sun. The birdwatcher has fewer worries about the wind direction than the student of mammals, because birds have generally little or no sense of smell, but gazing into a stiff breeze can be very eyewatering and irritating.

Rich may be the rewards of caution, quiet and stillness in woodland. Remain part-concealed for several minutes, and if no motion betrays your presence you may be accepted as just another part of the landscape, with birds resuming their normal lives around you. In this way, parties of birds like tits, Wrens, Treecreepers and Goldcrests may approach to within a few feet.

NOTEBOOKS AND RECORDING

We have already discussed the use of a camera to record particular bird habitats as an "extra" to your birdwatching, but a notebook to record details of numbers, plumage or behaviour is an essential. For over a century amateur birdwatchers have played a large part in the study of bird populations and their movements in Britain—and in essence this work is based on the facts and figures accumulated from scores of notebooks such as yours. This is the place to record your first Chiffchaff of the spring, or the last Swift or House Martin of autumn; to note the numbers of the various birds you see, especially in areas that you visit regularly. Also this is the place for the beginner to note details of new birds—how they strike him, how he feels that his bird-book let him down—so that he may identify this species with greater ease the next time he meets it. The same applies to the more experienced birdwatcher meeting a less-common species for the first time, and in either case, if the words can be accompanied by sketches, very much the better. Birders have coined the word "jizz" for that indefinable something that makes a bird recognisable—usually it is something to do with the way it stands, feeds or flies, rather than with its colouration—and this is often easier expressed in a sketch than in words. For a real rarity, such notes, taken in the field with the bird before you, are quite vital in deciding or confirming its identity.

Perhaps the best sort of notebook is the pocket-sized, loose-leaf variety, with a stiff waterproof cover. At the end of each year, the notes can be removed and filed (looking back on them later is immense fun—and often helpfully interesting), and a fresh series of pages inserted. The real record keepers, those with the tidy minds, find benefit in transferring all their records and notes for each species to a series of filing cards for quick reference. This is a habit that it is well to regard as a discipline, but (unlike many disciplines) it has obvious merit *and* interest, for from such records are county and regional bird reports compiled. Daily notes and sketches are perhaps best and most interestingly kept in diary form, but systematic records are best filed in a regular order. We saw earlier that taxonomists delight in changing the order in which the various species are listed—some being more *avant garde* than others. Regrettably, this only confuses the poor birdwatchers (as those who have tried to find their way about a new *Field Guide* and the old *Handbook* will well know). Recently the editors of a number of county bird reports got together to agree an order, and until this is replaced, it can sensibly be taken as the most practical. It is published (cheaply) by the B.T.O. as Field Guide No 13 under the title *A species list of British and Irish Birds*, and contains not only the accepted English names, but also the correct scientific ones. It is worth having a copy.

BIRD BOOKS

There is an incredible number of bird books—mostly useful, mostly interesting, some vital, some fascinating. They range from field guides to help you identify our birds in their various plumages, through encyclopaedic multi-volume sets, with details not only of plumage, but of anatomy, biology, behaviour and distribution, to monographs on a single species, regional bird books and travelogues.

It is, of course, impossible to list them all. What may be both possible, and sensible, is to indicate a few of the more important books for all birdwatchers and a selection of more general-interest topics. From this, the more dedicated bibliophile, or the budding ornithological librarian, will have found a starting point. The nature of a birdwatcher's library is essentially personal; both authors of this book, for example, have been involved with seabirds—for James Fisher books on seabirds formed only a section of his vast and comprehensive library, whereas for myself they dominate my small collection, indeed to the point of a nearly complete set of the works concerning St Kilda, a seabird Mecca.

New books are appearing at a tremendous rate, and it would be a brave man who set out to emulate the world-famous photographer Eric Hosking, who must now have one of the most complete modern private ornithological libraries. Thus many of the works mentioned by James Fisher in earlier editions of this book have been superseded—some, as we will see, have not, and remain in the "must" category. I have divided the list into books of various types, with here and there some notes. As with any other piece of equipment, only you can judge the book which is best suited to your working interest or taste. Most, if not all, of these books should be readily available in a public library, so have a look at them before plunging.

FIELD GUIDES TO IDENTIFICATION. Any birdwatcher is bound to need one or more of the comprehensive field guides in the course of time. Many more-elementary paperbacks on (for example) garden birds are available but are not listed.

The Birds of Britain and Europe with North Africa and the Middle East; Heinzel, Fitter and Parslow; Collins.

A Field Guide to the Birds of Britain and Europe; Peterson, Mountford and Hollom; Collins.

The Hamlyn Guide to the Birds of Britain and Europe; Bruun and Singer; Hamlyn.

HANDBOOKS. These are the really comprehensive works, with great and fascinating detail on all aspects of the birds included. The Witherby *Handbook* is the oldest, but remains the most comprehensive. Bannerman has a wealth of information, much of it personal or anecdotal. In preparation in several volumes is a new work, *Birds of the Western Palaearctic*, but this series is unlikely to be completed for several years yet. The two "Popular" volumes are cheaper condensations of the main work.

The Birds of the British Isles; Bannerman; 12 vols; Oliver and Boyd.

The Handbook of British Birds; Witherby, Jourdain, Ticehurst and Tucker; 5 vols; Witherby.

The Popular Handbook of British Birds; P. A. D. Hollom; Witherby.

The Popular Handbook of Rarer British Birds; P. A. D. Hollom; Witherby.

GENERAL INTEREST WORKS. Some of these (for example James Fisher's *Shell Bird Book*, David Lack's *Enjoying Ornithology* and Michael Lister's *Birdwatcher's Reference Book*) are excellent sources of further titles.

Breeding Birds of Britain and Ireland; J. Parslow; Poyser.

Atlas of European Birds; K. H. Voous; Nelson.

Bird; L. and L. Darling; Methuen.

A Bird and its Bush; M. Lister; Phoenix House.

Bird Navigation; G. V. T. Matthews; second edition (also in paperback), Cambridge University Press.

Birds and Woods; W. B. Yapp; Oxford University Press.

Birds of the World; O. Austin and A. Singer; Hamlyn.

The Birdwatcher's Reference Book; M. Lister; Phoenix House.

Enjoying Ornithology; D. Lack; Methuen.

An Eye for a Bird; E. Hosking; Hutchinson.

The Life of Birds; J. C. Welty; Constable.

Man and Birds; R. K. Murton; Collins, New Naturalist Series.

The Migration of Birds; J. Dorst; Heinemann.

Pesticides and Pollution; K. Mellanby; Collins, New Naturalist Series.

Population Studies of Birds; D. Lack; Oxford University Press.

Seabirds; J. Fisher and R. M. Lockley; Collins, New Naturalist Series.

The Shell Bird Book; J. Fisher; Ebury Press and Michael Joseph.

Where to Watch Birds; J. Gooders; André Deutsch.

Where to Watch Birds in Europe; J. Gooders; André Deutsch.

The World of Birds; J. Fisher and R. T. Peterson; Macdonald.

Bird Ringing; C. J. Mead; B.T.O. Field Guide.

TECHNICAL GUIDES. Helping with techniques or equipment.

Discovering Bird Watching; J. J. M. Flegg; Shire.

Binoculars, Cameras and Telescopes; J. J. M. Flegg, B.T.O. Field Guide.

The Bird Table Book; T. Soper; David and Charles.

Nestboxes; J. J. M. Flegg and D. E. Glue; B.T.O. Field Guide.

A Field Guide to Birds' Nests; B. Campbell, J. Ferguson-Lees; Constable.

MONOGRAPHS. This is a very brief selection from the dozens

available—many published under the auspices of Collins New Naturalist series.

The Greenshank; D. Nethersole-Thompson; Collins, New Naturalist Monograph 5.

The Hawfinch; G. Mountfort; Collins, New Naturalist Monograph 15.

The Herring Gull's World; N. Tinbergen; Collins, New Naturalist Monograph 9.

The House Sparrow; J. D. Summers-Smith; Collins, New Naturalist Monograph 19.

The Kingfisher; R. Eastman; Collins.

The Life of the Robin; D. Lack; Witherby.

The Mystery of the Flamingoes; L. Brown; Country Life.

Penguins; J. Sparks and T. Soper; David and Charles.

The Puffin; R. M. Lockley; Dent.

A Study of Blackbirds; D. W. Snow; Allen and Unwin.

Swifts in a Tower; D. Lack; Methuen.

The Woodpigeon; R. K. Murton; Collins, New Naturalist Monograph 20.

The Wren; E. A. Armstrong; Collins, New Naturalist Monograph 3.

RECORDS OF BIRD CALLS AND SONG

One of the areas in which experience can be of the greatest value to a birdwatcher (excepting those unfortunates who are completely tone-deaf) is in the recognition of birds by their calls or song. It really helps very greatly to be able to "spot" your bird in this way. Likewise one of the areas of greatest problem for the writer of a field guide is in "translating" a call, or, more difficult still, a song, into written English. Usually what results is a jumble of consonants looking like the name of a rural Polish railway station, and only rarely can the poor birdwatcher be significantly enlightened ("chiff-chaff" and "cuck-oo" for instance). There are other ways of learning, though, especially now that records, and record players are widely available. Here is a list of some record series that are available.

Listen the Birds; produced by the RSPB and the Dutch Society for Bird Protection; recordings by Hans A. Traber and John Kirby; a series (now fourteen in number) of 7-inch $33\frac{1}{3}$ r.p.m. discs, each with several related species, or with species common to one type of habitat; EPHT 11–14, and HDV 7–14.

Bird Recognition; an aural index recorded by V. C. W. Lewis on HMV; presented as three volumes, each with three 7-inch 45 r.p.m. discs and an explanatory booklet, again divided (with 15 or 16 species per volume) on a habitat basis; 7 EG 8923–31. Victor C. Lewis has also produced other 12-inch 33⅓ r.p.m. records for HMV; *A Tapestry of British Bird Song*, CLP 1723, sixty-odd species; *Guess the Birds*, XLP 50011, 24 species arranged as a test of skill as well as for entertainment; and, for Marble Arch, *Bird Sounds in Close-up*, two 12-inch records, MAL 1102 and 1316, both with a wide spectrum of about forty species arranged in habitat groupings.

Shell Nature Records; published by Discourses for Shell-Mex and B.P. Ltd; recorded by Lawrence Shove; this series is based on a habitat classification—marsh and riverside, moor and heath, woodland, etc.—with seven 7-inch 33⅓ r.p.m. records so far; DCL 701–7.

BBC Records; the Wildlife Series contains bird recordings, again divided on a habitat basis; RED 96 has Welsh birds, RED 103 a large number of woodland birds, summer and winter, RED 109 garden and water birds, etc. 12-inch 33⅓ r.p.m.

It is well worth "getting into practice" for the forthcoming season by, for example, spending the occasional March evening listening to recordings of the summer migrants that will soon be arriving. Likewise, you may search your "woodland" records in the hope of identifying the call of some unseen bird (though most of these untraceable calls turn out to have been made by the very versatile Great Tit!). Many difficult, or distantly-seen species like waders, become much easier to identify once you are familiar with their calls.

One step beyond this, and regarded by more and more birdwatchers as another tool of their trade, lies the portable tape recorder. Recording equipment—parabolic reflectors and all the other paraphernalia—is outside the scope of this book, but it is worth bearing in mind that a tape recorder can be used not only to record bird-song spectaculars, such as the Nightingale or dawn chorus, but also to draw birds towards you. Many birds (but some have shown themselves aloof to temptation so far) will come to investigate the source of a "broadcast" of a recording of their call or song—often, like the Nightingale, coming right out into the open to pour forth a reply. Here is yet another technique—novel, and ripe for much more development—for getting those close views that enhance your birdwatching so much.

4 Migration

"We must not, I think, deny migration in general: because migration certainly does subsist in certain places, as my brother in Andalusia has fully informed me. Of the motions of these birds he has ocular demonstrations."

GILBERT WHITE, *12 February 1772*

In practically every group of animals, wherever these animals live, there can be found members which indulge in orderly mass movements. These movements may happen every few hours, every day, every few days, every month, every year, or every few years. Animals do not, as far as we know, have regular movements of the order of every week, because the week is a human invention which bears no relation to natural phenomena like the daily turning of the earth, the transit of the moon, or the movement of the earth round the sun.

The lives of most birds are based on an annual rhythm; although in some large birds, particularly tropical seabirds, this may not hold true. This annual rhythm may produce several sorts of migration. It may consist of wholly local movements, or dispersals in no particular direction. It may consist of the typical more or less south to north movement that makes it possible for us to label such a large proportion of the birds on the British list as summer or winter visitors. This south to north movement may even involve crossing the Equator. In the tropics, however, there may be orderly seasonal movements, but these may not be concerned at all with latitude. The birds move every year at the best time to avoid the dry season, and these movements every year may be quite irregular in their nature and yet orderly because they always happen at the same season. When we discuss the special case of rhythms extending not over one but over several years, we find ourselves dealing with the remarkable cases of periodic irruptions, movements of birds when their population is high into geographical areas not usually associated within their natural range.

LOCAL MOVEMENTS

Many kinds of birds are described as residents or sedentary, because they are never known to move very far from where they are hatched. Yet every year these birds move about locally and sometimes this movement is more or less orderly. Some birds move up and down mountains. Some form flocks in winter and leave their breeding territories to roam together about the hedges and fields of the countryside in search of food. The Yellowhammer and Chaffinch are typical British birds which never appear to migrate in the true sense, but which flock up and roam about open country. Similarly tits and Goldcrests, highly resident birds in Britain, flock up in autumn and go on foraging parties for the winter. Of course, the picture is somewhat clouded in Britain and Ireland in winter by the large numbers of some species (like the Chaffinch and Goldcrest) which arrive here as migrants driven out of the Continent by the severe winter.

There seems to be scope for a fair amount of individual variation, since some individuals of normally resident species may carry out true migrations. For instance, the Starlings that breed in Britain are almost certainly resident and do not move more than locally, whereas those in the neighbouring part of the Continent truly migrate, in a more or less east to west direction, so that they may arrive in Britain in the winter.

Local movements start rather earlier than most people imagine, since even before the parent birds have finished with their last brood, the offspring of the first broods may have been driven away and may have scattered through the country. These young birds often join with the flocks and quite often roost in company. The early autumn movements seem quite highly organised and deserve a good deal more study than they have had so far.

DISPERSALS

The movements of many birds are governed by the necessity for breeding in the summer season, rather than any particular need to move southwards in winter. Yet the movements of these birds are too large and too definite to be called local movements. This especially applies to sea-birds, which may disperse for many hundreds and sometimes even thousands of miles from their breeding haunts. For example, the two young Puffins marked on St Kilda, the most westerly of the Scottish islands, which had crossed the Atlantic to be recovered in Newfoundland. They must have done this as much by swimming as by flying. The distance was over 2,000 miles.

Sea-birds have the whole of the ocean to feed in, but only a limited number of sites to nest on. Dispersal is therefore very wide, and Gannets, which only breed in a handful of colonies in Britain spread out all over the Continental shelf, where the water is not too deep, in the winter. The direction of their spread may be north, west, east, or south; where the bird goes seems to depend largely on its individual inclinations. This is not quite true of young Gannets, which seem, during the three years before they can breed, to indulge in true north and south migration, reaching the coast of North-west Africa.

Kittiwakes, Fulmars and other petrels, and most of the auk tribe are oceanic birds in winter. Their spread from their breeding haunts may indeed lead them south in the Arctic regions, where ice occupies the open sea; but where there is no ice their movements seem to be in the direction of adequate food supplies, or in no particular direction at all. Besides ice, one of the factors limiting where they can go is probably heat tolerance. Thus the Fulmar goes as far south, summer or winter, as the 60°F isotherm; which means that it goes farther south in summer than in winter. This does not mean, however, that the Fulmar is an ordinary north to south migrant. Its movements are far more accurately described as dispersals.

MOVEMENTS AT SEVERAL-YEARLY INTERVALS

It is becoming clear that a far larger number of birds than was at first supposed undergo more or less regular changes in population. When a peak of population is reached, it very often means that the pressure of numbers causes a big movement which it seems reasonable enough to call a migration. Some authorities describe these as irregular migrations or irruptions. Perhaps the latter term is better, since one of the characteristics of these mass movements is that the interval between them is often much the same, between nine and twelve years, and almost always between five and sixteen.

In Britain irruptions of the Crossbill arrive at quite short intervals (every three to ten years), and in some years these have been so large that a remnant has actually stayed on to breed here. The permanent colony of Crossbills in Norfolk was established after one of these irruptions. Pallas's Sand-grouse used to appear in Britain in large numbers about every twelve years, but has not done so now for fifty or sixty years. Rose-coloured Starlings seem to have a cycle of nine years or so, and at the peak of this often irrupt from Asia to Europe and breed for a year or two. In these peak years birds quite often reach Britain.

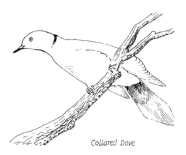

Collared Dove

RANGE EXTENSIONS

Although perhaps not strictly falling under the general heading of migration, this is a good place to mention the fantastic expansion of range that the Collared Dove has undergone in recent years. When James Fisher wrote the earlier editions of this book, he made no mention of the Collared Dove, which was then considered, if at all, as a rather drab relative of our Turtle Dove, resident in Turkey and other areas in the eastern Mediterranean. In the years following the Second World War, this previously rather static species became suddenly dynamic, and started to spread rapidly westward across Europe. In 1955 and 1956, the first two or three pairs reached eastern England to breed (shrouded in secrecy and excitement). As I write (in 1973), the Collared Dove breeds in all, or almost all, British and Irish counties, and on many of the western Irish and Scottish islands. Even on remote St Kilda, where there is little cover for it to breed, it is a regular visitor (perhaps even a non-breeding resident) and two years ago its irrepressible westward push resulted in breeding pairs in Iceland! There must now be tens of thousands of breeding pairs in the British Isles—from only one pair less than twenty years ago. Indeed in some parts of eastern England (and even around grain silos in western Ireland) many people consider Collared Dove numbers and feeding habits sufficiently in conflict with man to call them "pests".

TYPICAL MIGRATION

Most of our summer visitors and nearly all our winter visitors in Britain are typical migrants, that is, they are birds which have two distinct ranges, though in many cases these ranges may overlap. Between these ranges they have an orderly movement in spring and autumn. When they are making this movement we say they are on passage.

What, then, are the reasons for these regular seasonal migrations? Primarily, we must suppose, creatures endowed with the power of flight have the opportunity to exploit areas which, while rich in food in the summer months, are impossibly cold in winter. Obviously there are degrees in this. Thus some species (like our summer visitors) move only

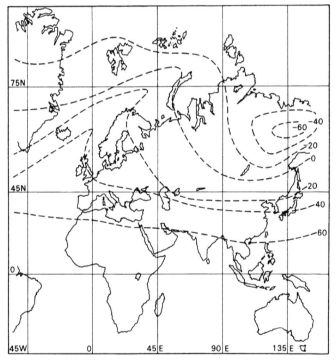

17   Isotherm map showing mean January temperatures (F)—note cold centre in eastern Siberia

from overwintering areas in the tropics to our temperate climates, and may fit in two or three broods of young during their visit, while others (like many ducks, geese and waders) overwinter in temperate regions like western Europe before hastening northwards to take advantage of the very short Arctic summer. They must arrive on the tundra as the snow and ice retreats, for summer is short. However the "crop" of insects waiting to be harvested is enormous in a favourable year, and rearing a single brood may not be difficult if the timing is right. The Sanderling—a wader a little smaller than a Song Thrush—has evolved a neat additional technique for these circumstances. On arrival in the Arctic, the pair build two nests, and the female lays a clutch of eggs in each. She then sits on one clutch while the male incubates the other—thus raising two broods in the time normally taken for one.

A glance at the map of the mean January isotherm shows clearly that much of the Palaearctic region (comprising Europe and Asia) is completely inhospitable in winter. Clearly, too, if birds retreat from the cold following the temperature gradient, many will move west as well as south (the Himalayas provide another incentive for doing this) to the coasts of Europe warmed by the Gulf Stream, which extends a finger well to the north of Scotland. Hence the offshore islands of Britain and Ireland are a winter refuge for huge numbers of a wide variety of species (Fig. 17).

It is often said that Britain is well situated for students of bird migration, but in fact there may be so much movement, of so many types, that the picture becomes rather clouded. There is a stream of migrants in autumn from Greenland and Iceland—some passing on (Greenland Wheatears), others staying (Barnacle Goose). Another stream leaves the Arctic via Scandinavia, again with some species staying with us (Knot and Dunlin) and others (Arctic Tern) moving on. The Arctic Tern probably spends more time in daylight than any other living creature, for it breeds in the perpetual days of the Arctic summer, migrating south beyond Africa to "winter" in the perpetual days of the Antarctic summer on the fringes of the pack ice.

Add to these movements the departures of our Swifts, Swallows, warblers and all the rest of our summer visitors, and the arrivals (timing and quantity depending on how hard a winter it is) of all the weather-displaced birds from central Europe, and the picture really does become complex. Nor is it a "once-and-for-all" movement, for should the weather worsen (or improve) during the winter, additional traffic will arise as flocks of birds are affected. The most characteristic examples of this weather movement, to be seen each year in Britain and Ireland, are Skylarks, Lapwings and Woodpigeons. Westerly-moving flocks of any of these species warn the birdwatcher to stoke up his fires!

Now we have some ideas of the benefits of migration to birds in allowing the colonisation, even temporarily, of large areas of the earth's surface, what are the disadvantages, the hazards of the immense journeys that may be involved? Clearly predators will be waiting at every turn. The toll taken of migrant songbirds by man—for delicatessen use, not to satisfy real hunger—is estimated at many millions of birds each autumn at the hands of the bird-limers of the Mediterranean alone. In Africa, one of the falcons has reversed its breeding season, raising its young late in the summer so that the parents have easy and profitable feeding on the streams of migrants passing by southwards.

Weather, too, must pose problems. Storms can literally knock small birds from the air, fog and rain can confuse their otherwise brilliantly accurate navigation. (It is always sobering to remember that in seven grammes of Chiffchaff is centred a navigational computer capable of guiding the bird to not only the same country, but to the same county and the same copse, when it returns next season. All without the aid of maps and the aids that space-age man finds essential!) Headwinds can increase exhaustion, and if a sea or desert crossing is involved, this could be fatal.

Sedge warbler

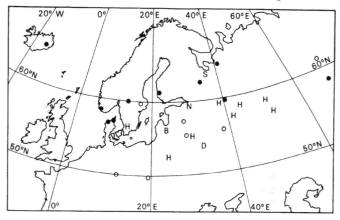

18 Lapwing—recoveries during the breeding season of birds ringed as young in Belgium= B, Denmark= D, Holland= H, Norway= N, Britain= O, Sweden= S

Just as astonishing as the navigational ability is the mechanical ability of the bird as a flying machine. "Fuel" for the journey must be carried—usually in the form of fat deposits in the body cavity and under the skin. The Sedge Warbler, for example, weighing 12 grammes or so in normal trim, may feed-up to such an extent before leaving Britain in autumn that it weighs more than 20 grammes—appearing almost spherical. In this state it takes off on the journey south carrying, it has been estimated, fuel enough for between 60 and 90 hours flying—certainly enough to reach Africa in one hop!

STUDY TECHNIQUES

Just how do birdwatchers investigate these movements of birds, perhaps not unjustly described as the miracle of migration. Since Solomon—"and the voice of the turtle (dove) is heard in our land"—man has been aware of, and fascinated by the topic. Even so, only a couple of hundred years ago, Swallows and martins were thought to hibernate in the mud at the bottom of ponds! At the turn of the century the technique of marking birds with a numbered metal ring was developed, and since then, with a host of amateur birdwatchers taking

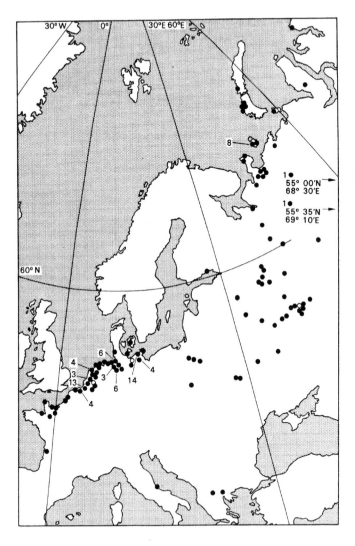

19   Whitefront—recoveries of birds ringed in Britain

part, our understanding of the routes, timing and destination of migrations has increased immensely.

Each ring (or "band" in America) has a serial number, individually identifiable just as is a car number plate, together with the return address of the ringing centre. In Britain, the national ringing scheme is administered by the British Trust for Ornithology from their headquarters at Tring, in Hertfordshire, and rings are marked either with "BTO, Tring, England" or "British Museum (Nat. Hist.) London SW7". The second address is most commonly used, and is thought to be more satisfactory because of the international understanding of words like "museum" and "London". Details of ringing—the species, its age and sex, date and place of ringing, are sent to the central files by the ringer, so that when anyone, at home or overseas, finds a ring, the original details can be linked with the finding details, the finder informed, and the "recovery", as it is now called, stored in the data bank—whence it can be withdrawn to help produce maps such as Figures 18, 19, 20 and 21. Should you find a ringed bird, send all details about it to the address you find on the ring, or (even if it is from a foreign ringing scheme) to the B.T.O., who will see that it speedily reaches the correct destination.

The rings themselves (Plate 13) come in many sizes, closely tailored to fit our birds. The smaller ones are made of lightweight aluminium alloys and weigh about 600 to the ounce (about as much of a burden to a bird as your watch is to you). For larger birds, especially seabirds, these alloy rings wear out before the bird does (many seabirds live for twenty years or more), and specially strong, corrosion-resistant nickel alloys have been developed for these conditions.

Naturally the fitting of the ring is a delicate job, requiring a considerable degree of skill and care in handling the birds. In Britain, ringers must have a licence issued by the Nature Conservancy Council (a Government Department) as well as a permit from the B.T.O. The issue of a permit is only allowed after extensive and detailed training in all aspects of bird handling—usually given at special courses, at bird observatories, or by selected highly-experienced ringers. Details are available from the Trust. At the moment, there are about 1300 ringers in Britain, who between them mark around half a million birds each year. From these ringed birds, we expect to hear again of something like 14,000 each year, recoveries coming in to Tring from all quarters of the globe (Fig. 22).

Many birds are marked as youngsters, still in the nest, which is es-

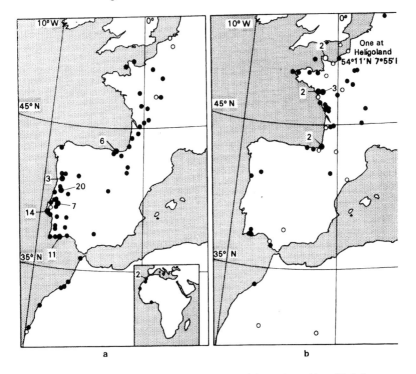

20   Recoveries of birds ringed in Britain—(a) Reed Warbler; (b) Sedge
Warbler

pecially valuable as they are of precisely known age and origin. The
majority, however, are caught for marking by a variety of techniques,
mostly developed from those used by hunters catching birds for food.
They range (Fig. 23) from the huge Heligoland traps used at observatories to the stick, sieve and string that can be used in any garden. For
ready portability, very fine nets (about the texture of a hairnet, some 9
feet high and from 20 to 60 feet long) called mist nets, originating in
Japan, are unbeatable. Their catching efficiency, erected against a
background of bushes or trees, is very high because they are fine enough
to be invisible to all intents and purposes to the flying bird. So soft and

elastic are they that the flying bird is brought gently to rest before drop-
ping into a pocket, unharmed, to await removal by the ringer (Plate 3).
Bigger birds, such as ducks and geese, can be caught in a more robust
net, which is fired quickly and safely over their heads by means of
rockets or "cannons". This technique has recently been developed to a
very high degree; and amazingly, for some populations of these birds,
between 1 in 10 and 1 in 50 of all the individuals in the world carry a
ring.

DIRECT OBSERVATIONS

Ringing is not the only way of studying birds migration. Hundreds of
birdwatchers all over Europe contributed records of their Swallow
sightings in spring so that H. N. Southern's map (Fig. 24) of arrival
dates could be compiled. Again, scattered over Europe is a network of
bird observatories, all situated in areas where experience has shown

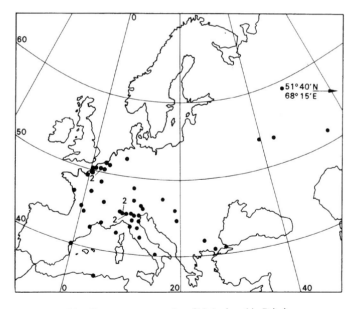

21   Garganey—recoveries of birds ringed in Britain

22   Ringing recoveries for some of the bird species ringed in Britain—note the Arctic Terns off S.E. Australia, more than 10,000 miles (17,000 km) from the point of ringing

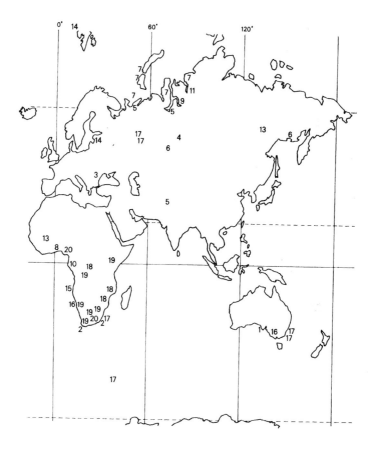

passage movements to be both regular and heavy. Some are on islands (like Heligoland and Fair Isle) and many are on the coast. Still others are situated inland, like the Col de Bretolet high in an alpine pass, or Radolfzell, on a peninsula in Lake Constance in Germany. Britain has a

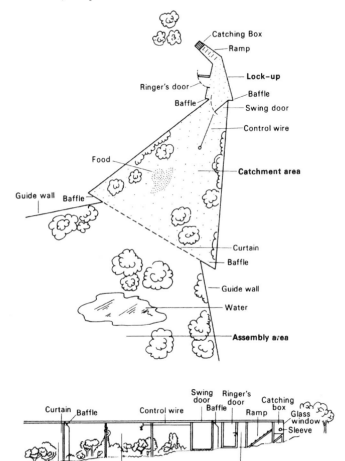

23 The Heligoland trap—note the baffles and the funnel effect of the guide walls to the assembly and catchment areas

good number of observatories, mapped in Fig. 25, where generally there is a full-time professional warden, who is helped in his observations, counting, trapping and ringing by teams of volunteers who may visit the observatory for a few days or weeks in the spring or autumn migration seasons.

Many birds, however, migrate through the night, and such "visual observation" techniques cannot be applied. However, during the Second World War and afterwards, the scientists who were developing radar were puzzled by slow-moving belts of interference, looking rather like a snowfall across the screen. The name "angels" was coined for this interference, which was eventually found to be due to flocks of migrating birds. While angels caused problems to military and civil aviation controllers, ornithologists (under the leadership of David Lack in Britain) saw in them an ideal method of watching migration after dark, or when it was too high to be detected by visual observation.

One of the major findings from radar studies showed just how much migration takes place on broad fronts. It had been thought that migration routes generally followed quite narrow "channels" along leading lines such as coasts, rivers or hills. Sometimes this is still the case, for example when storks and birds of prey, which depend on hot-air currents rising off the land for migration, cross the Mediterranean at Gibraltar or the Bosphorus, the shortest sea crossing. Generally, even so, the nocturnal arrivals of migrants recorded from radar observations are on fronts usually tens and often hundreds of miles wide. Problems still remain, however, the main one being that the "blips"—the echoes left on a radar screen by birds—cannot be identified as to species. Sometimes it is possible to say that they are Buzzard-sized, or Lapwing-sized, but rarely more. One species is characteristic: the Starling. The echoes of Starlings leaving their roosts in the early morning are called "ring angels"—and look just like the concentric ripples that spread out after you drop a pebble in a pond. These ring angels have shown Starlings leaving in all directions from their roost, in batches, spaced out by very regular time intervals.

For small birds, it is not possible to estimate with much accuracy how many birds, in a small flock, produce each blip, but the sight of a radar screen, covering perhaps a circle 100 miles in diameter, wholly obscures by many thousands of echoes, gives some idea of the really vast numbers of birds involved in migratory movements.

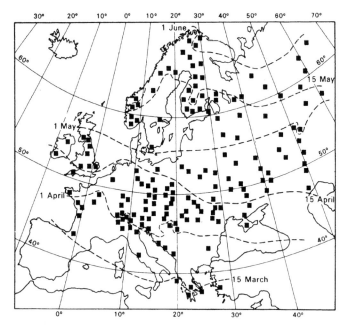

24 Swallow sightings in spring, showing the rate of spread of arrivals. The dotted line indicates the approximate position of the main advancing front on the dates given (after H. N. Southern)

OTHER USES OF RINGING

For over half a century, ringing recoveries have been accumulating. For some species we now have a fair idea of the pattern of movement, for other perhaps less common species, this picture is only just emerging. For example, it begins to look as if the Lesser Whitethroat, a charming and inconspicuous hedgerow and scrub bird, migrates southwards in autumn to northeast African countries such as the Sudan *via* Italy. The return journey in spring is made well to the east of this route, through Syria and the Lebanon, and thence north of the Alps to Europe. The explanation of this "loop" migration is still being sought. Even for those species that we think we understand, we must not relax,

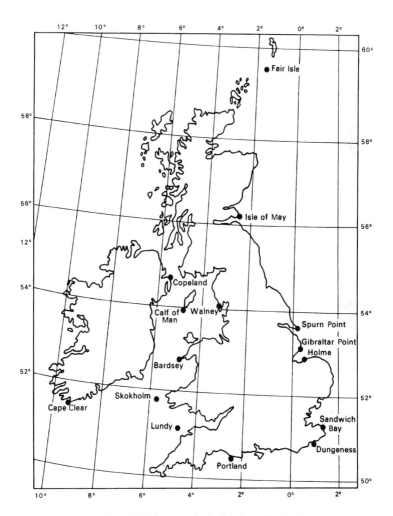

25 Bird observatories in Britain and Ireland

for bird populations are dynamic things and there are (for example) recent indications that Blackbirds may be breeding further north in Scandinavia and wintering further south in Iberia.

I have already indicated that the time elapsed between ringing and recovery will give some idea of the life span of the bird. The table below lists the known life spans for a variety of species:

TABLE OF LONGEVITY

Species	Oldest	Species	Oldest
Storm Petrel	15.0	Swallow	16.0
Manx Shearwater	20.0	Raven	12.5
Fulmar	22.0	Rook	15.9
Gannet	12.11	Jackdaw	14.0
Cormorant	19.7	Magpie	9.1
Heron	18.0	Jay	15.9
Wigeon	18.3	Great Tit	7.0
Pintail	16.8	Blue Tit	11.3
Tufted Duck	13.11	Coal Tit	7.4
Eider	15.7	Long-tailed Tit	5.5
White-fronted Goose	17.8	Treecreeper	7.2
Pink-footed Goose	16.2	Wren	4.11
Mute Swan	14.4	Mistle Thrush	7.8
Kestrel	14.5	Song Thrush	10.2
Moorhen	11.4	Blackbird	11.0
Oystercatcher	20.4	Nightingale	4.1
Grey Plover	11.11	Robin	8.4
Woodcock	12.4	Reed Warbler	10.1
Curlew	16.6	Sedge Warbler	6.1
Redshank	14.5	Blackcap	6.1
Knot	13.0	Garden Warbler	5.8
Dunlin	11.11	Whitethroat	6.4
Arctic Skua	16.10	Willow Warbler	5.9
Herring Gull	16.1	Chiffchaff	3.7
Black-backed Gull	19.11	Dunnock	9.6
Common Tern	25.0	Pied Wagtail	9.11
Arctic Tern	27.0	Starling	13.10
Razorbill	20.4	Greenfinch	11.0
Puffin	15.9	Goldfinch	6.8
Woodpigeon	15.11	Linnet	8.1
Turtle Dove	10.8	Bullfinch	6.7
Cuckoo	4.0	Chaffinch	10.7
Barn Owl	9.0	Yellowhammer	8.9
Tawny Owl	19.0	House Sparrow	10.6
Swift	15.11	Tree Sparrow	6.7
Kingfisher	2.10		

Similarly, we can derive some idea of the causes of death of those birds reported to us, even though they tend to be rather weighted towards birds that have died in proximity to man. In this way the magnitude of the effects of the wreck of the tanker *Torrey Canyon* and the subsequent oiling disaster in 1967 could be traced, from recoveries of auks from all the Irish Sea colonies involved. In the past, hundreds of British-ringed finches—Linnets, Goldfinches, Chaffinches and Red-polls especially—were caught and caged by Belgian bird-fanciers. Such figures were used as evidence in the attempts to restrict or, better, eliminate this sport in the interests of the conservation of these species. Likewise, for some of our summer visiting warblers (such as the Blackcap), as many as 25% of the recoveries we receive each year are reported from the bird-catching regions of the Mediterranean. Of the very few young Ospreys reared in Scotland under the costly protection of the R.S.P.B. and other bodies, at least four have now met their ends in the Mediterranean, shot by "hunters".

It would be quite wrong to close this chapter on migration without emphasising one major point: despite the obviously giant strides we have made since Gilbert White's day, we still have no complete solution to the mystery of migration. How bird navigation works in so many different circumstances must remain the greatest unsolved biological problem of our time. There are many theories—too complex to be dis-cussed here—each with some experimental evidence behind it, but for a full solution, an explanation of the necessary brain structures, and for the accuracy of time, distance and direction judgements, we must wait.

Lesser Whitethroat

5 Where birds live

"Selborne parish alone can and has exhibited at times more than half the
birds that are ever seen in all Sweden; the former has produced more
than one hundred and twenty species, the latter only two hundred and
twenty-one. Let me add also that it has shown near half the species that
were ever known in Great Britain."

GILBERT WHITE, 2 September 1774

Most creatures, except the very unspecialised and adaptable ones such
as man, are governed to a remarkable extent by the nature of their environment. Birds are no exception. A species of bird out of its
geographical range or its habitat is usually an oddity.

This book is written with special reference to the islands of Great Britain and Ireland, and it might be as well, therefore, to describe this part
of the world from the bird's point of view. The British Isles are rich in
bird life. From the point of view of animal geography, they lie at the
edge of one of the most important of the great zoological regions—the
Palaearctic. This zone extends from Britain to Japan, and its animal
(and bird) inhabitants are in the mass different from those in the other
parts of the world. The great region of the Nearctic, which comprises
the north of the American continent, has the closest affinities with the
Palaearctic region, but in many respects the birds, particularly those of
the land, are different.

Of the countries in the Palaearctic region Britain has probably a
larger list of birds than any other of like area and climate. This is largely
due to the facts that it lies athwart the west coast of Europe, directly in
the path of a major migration route; that it is provided with many
diverse habitats—large areas of undeveloped marshland alone being
absent; that its northern oceanic cliff-bound coasts attract sea-birds and
others with arctic affinities; and—by no means least—that bird-
protection and conservation (with one or two unfortunate exceptions)
are almost second nature to the modern British, and that only a small
group of species is slaughtered for food.

From the point of view of climate, Britain is more favourable to birds than many other parts of the Palaearctic region (Fig. 17). Though situated in a rather northerly latitude, it is kept warm in summer and merely cool in winter, by the Gulf Stream; the atmosphere is not too dry, not too moist; there are no violent extremes of drought or rainfall. As a result, it has a large population of resident birds, a great number of visitors which reach it in summer from tropical countries, and many which visit in winter from the arctic (for even the arctic is cool in summer, and our winter climate is little different from this) and also from parts of central and eastern Europe, which although on similar latitudes to Britain, have winters of continental severity (20°C or more colder than ours!). The great number which visit our country on passage, in autumn and spring, do so because Britain offers a reliable and obvious migration route.

Very little of the surface of Britain is as it was before civilisation came. In the north conifers and birch have given way to moorland, in the south oak scrub or forest to parkland, plantations, pasture, and crops. And it is the surface that matters to the birds. In historical times many species must have come and many have gone, and many have changed their habits. The following table shows the main features of the plant-environment in which British birds live.

A. WOODLAND:
    (1) Coniferous:
        (a) Pine (natural).
        (b) Plantations.
    (2) Deciduous:
        (a) Birchwood.
        (b) Oakwood—pure or mixed (sandy soil) and oak-birch heath.
        (c) Beechwood—on loam (often with oak or ash) or on sand.
        (d) Ashwood, often with oak, sometimes birch.
        (e) Alder, sometimes with birch and willow species.
        (f) Scrub, often hazel, ash or hawthorn.
    (3) Mixed:
        (a) Plantations, various types.
        (b) Beech on chalk with yew and ash.
        (c) Birch with juniper.

Most of the above types can be planted or natural; with open or closed canopies; with or without secondary growth.

B. PARK OR GARDEN LAND:
   (4) Fringing woods and coppices.
   (5) Parkland.
   (6) Orchards.
   (7) Gardens.
C. AGRICULTURAL LAND:
   (8) Allotments, smallholdings, nurseries, etc.
   (9) Arable farmland.
   (10) Grassland.
D. HEATH AND MOOR:
   (11) (a) Lowland (rough pasture, commons)
        (b) Acid heaths with gorse.
   (12) Upland (sometimes rough pasture).
E. ALPINE:
   (13) Mountain tops.
F. WATER LAND:
   (14) Flowing.
        (a) Streams.
        (b) Rivers.
   (15) Placid:
        (a) Ponds and small lakes.
        (b) Large lakes.
   (16) Stagnant:
        (a) Mosses and bogs.
        (b) Marshes and fens.
G. COASTLAND:
   (17) Salt Marsh and estuary.
   (18) Dunes.
   (19) Beaches.
   (20) Cliffs.
H. MAN-LAND:
   (21) Built-up areas.

It must be stressed that this is a general classification of the habitats of Britain arranged as far as possible from the bird's point of view. Of course there are many other kinds of habitat in the world with many other subdivisions. In the tropical forests alone it is possible to describe about eight layers, one on top of the other, with a special bird fauna to each. But such extensions of habitats need not concern us here. Our

1 A Puffin from a specially marked population on St Kilda. The numbered B.T.O. ring is on its left leg, and a coloured plastic ring, with etched letters LULU, is on the right—the lettering can be read through binoculars at a considerable distance. With rings such as these the diminishing population can be more easily studied in an attempt to understand the causes of its sad and rapid decline. (Jim Flegg)

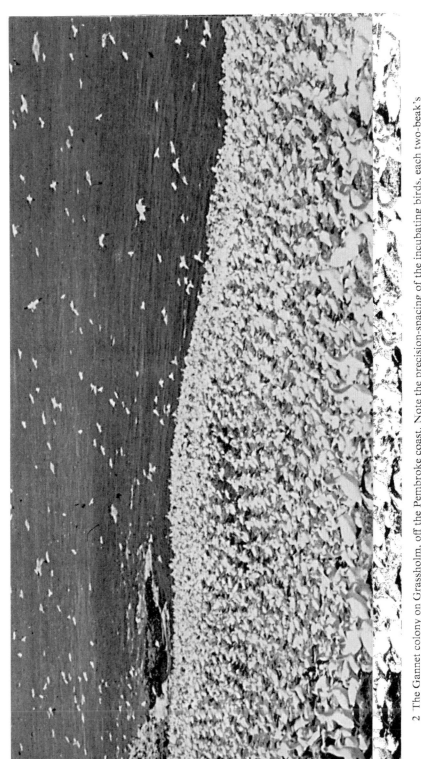

2 The Gannet colony on Grassholm, off the Pembroke coast. Note the precision-spacing of the incubating birds, each two-beak's distance from its neighbour. (Photo: Jim Flegg)

3 A line of mist nets set in a woodland clearing. Note the net in the left foreground, which stretches away well beyond the ringers. (Photo: Jim Flegg)

4 A chardonneret garden trap which has caught a Greenfinch coming to peanut bait. The other cell remains set and the trapped bird may tempt another to enter. (Photo: Jim Flegg)

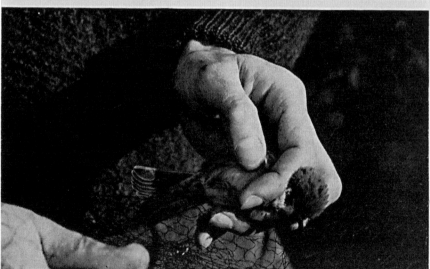

5 The skilful and quite gentle process of extracting a Tree Sparrow from the mist net. (Photo: Jim Flegg)

6 The ringer now has the bird comfortably held in such a way that it cannot struggle. (Photo: Jim Flegg)

7 Applying the ring, with special pliers, to a Common Sandpiper. For wading birds, strong alloys are necessary to prevent the ring wearing out. (Photo: Jim Flegg)

8 A special wedge-shaped nestbox for Treecreepers. (Photo: Jim Flegg)

9 The birdwatcher has great opportunities to experiment in tempting unusual species into his nestboxes. Here, open-fronted boxes have been set at a sewage disposal works to attract Pied Wagtails. (Photo: Jim Flegg)

10 A male Robin attacks the red breast of a stuffed bird hanging from a perch. The Robin is one of the most aggressively territorial of British birds. (Photo: Richard T. Mills)

11 By gently blowing aside the feathers, a ringer exposes the fleshy 'brood patch', well supplied with blood vessels, with which the female Hawfinch (and other species) warms her eggs during incubation. (Photo: Jim Flegg)

12 Ringed Plover about to settle on her pebble-shaped, pebble-coloured egg. Note how the belly feathers are fluffed out so that the 'brood patch' will be in contact with the eggs. (Photo: Richard T. Mills)

MUS. NAT.
REYKJAVIK 26584
ICELAND

ST. ORNITH.
POLONIA VARSOVIA
E-1040499

VOGELTREKSTATIE
ARNHEM-HOLLAND
1088915

CSSSR EKRB BUSSR
MOSKWA &-8 8 2 ss1

9508308
RIKSMUSEUM
STOCKHOLM

MOSKWA
78489   E

ESTONIA-MATSALU
-17 886

BRIT. MUSEUM
LONDON SW7
5062072

KITTIWAKES (6 years)

DS33154

SANDWICH TERN (8)

374191

STARLINGS

STORM

PETREL

RAZORBILL

13 Bird rings of various sizes from several countries. Note how worn the tern and petrel rings have become. (Photo: Jim Flegg)

14 Common Tern chicks, showing the camouflage pattern of their backs which conceals them as they crouch, and the white 'bib' so conspicuous when they raise their heads. (Photo: Jim Flegg)

task is to describe the places birds inhabit in temperate Britain and the kinds of birds which live in such places. The picture we get will at the same time give us a good working idea of the situation in North America, because there the places in nature are roughly the same, and though the birds that fill them may be of different species, they do much the same biological work.

To take our first category, we find for instance that the coniferous forests of America, Europe, and Britain (the last relic of primeval pine forest is in the Spey Valley) have birds breeding in them which are strikingly the same, as far as their adaptations to habitat go. A list might run:

American forests	*European forests*	*British forests*
(1) Wood Grouse	Willow Grouse Capercaillie	Capercaillie
(2) Grosbeaks	Grosbeaks	
(3) Siskins	Siskin (different species)	Siskin (as European)
(4) White-winged Crossbill	Continental Crossbill, Parrot Crossbill	Scottish Crossbill
(5) Spotted Woodpeckers	Spotted Woodpeckers	Spotted Woodpeckers
(6) Nutcracker	Nutcracker (different species)	
(7) Jay	Jay (different species	Jay (as European, but subspecies).
(8) Chickadee	Continental Crested Tit	Scottish Crested Tit Goldcrest and
(9) Kinglet	Goldcrest and Firecrest	Firecrest

The bird population of conifers in Britain is rather interesting because, except for those in the Spey Valley, practically all our coniferous woods belong to artificial plantations. After the last Ice Age, perhaps 15,000 years ago, it is probable that pine and birch forest extended well down to the south of Britain. The birds that lived in those woods originally spread down with them. Lack and Venables suggest that they were the Crossbill, Siskin, Lesser Redpoll, Treecreeper, Goldcrest, Coal Tit, Willow Tit, Crested Tit, Capercaillie, and Black Grouse.

Later the pine and birch woodlands moved north again (at the dawn of the Christian era it is probable that the only pine forests were within the Scottish Highlands), to be replaced by the typical oak forest of the south, and the typical conifer birds were left with two alternatives—to retreat with conifers or to alter their habits and colonise the oak. Crossbills and Crested Tits and Black Grouse, among others retreated, and the first two are now relics only in Scotland (the Crossbills that now breed in conifer plantations in south England belong to the Continental race and have come over in irruption years). Treecreepers, Coal Tits, and Goldcrests, however, have carved out a niche for themselves with some success in the oak woods; they also are typical birds, of course, of the southern conifer plantations which have been made by man in recent times. And into these planted pinewoods have also spread some birds of the broad-leaved woods; these plantations are seldom allowed to grow old, and in their early stages they are colonised by warblers, Song Thrushes, Robins, Wrens, and Dunnocks and some exciting species like Hen Harriers.

Just as there is very little coniferous woodland left native to Britain, so there is very little deciduous. Practically all the woodland you see in the country today is planted. Yet whether it is planted or not, we can recognise many types, each having a fairly distinct bird fauna. An important element of deciduous woodland is what can be generally called Northern Scrub. It is often composed of alder, often with birch and many of the different species of willow, and in the untouched northern regions it forms a fringe between the edge of the coniferous forests and the real tundra. Iceland and Scandinavia have much of this scrub, but in Britain it has nearly all been cut or grazed away. The typical birds of this northern scrub-land are finches like Redpolls and thrushes like Redwings. There are still plenty of Redpolls nesting in Britain, indeed they are presently (1973) increasing rapidly (Fig. 26), but with the loss and cutting away of most of their proper scrub-land, the Redwings have gone (probably some hundreds of years ago). Today the discovery of a Redwing nesting in Scotland is usually kept confidential although both they, and Fieldfares, are spreading very satisfactorily. They are most welcome re-colonists.

The main advantage which deciduous woodland has over coniferous is, from the bird's point of view, the presence of numerous holes in which to nest. The anatomy of a broad-leafed tree tends to permit the development of holes, whereas that of a narrow-leafed tree almost

26   Redpoll—the rapid increase of this species nesting in recent years, using 1966 as the index (100), is shown by this graph prepared from CBC (Common Bird Census) reports, published annually in the BTO's journal *Bird Study*.

prohibits this, especially in commercially managed plantations with an emphasis on so-called forest "hygiene", where all dead wood is removed. Often the number of species in conifers is less than in deciduous woodland, especially if the trees are really densely planted. Sometimes the total number of birds may be less, but by no means as often as people think. Conifers do not produce the terrific (but short-lived) peak "crop" of caterpillars associated with oaks, which are used with superb timing by the Blue Tit and Great Tit to feed their single, large brood (often between 10 and 15) of young, for example. Insect productivity in conifers is spread over the whole summer and benefits species with several broods of young.

Mixed woods, plantations of perhaps beech and pine, spruce and oak, can be very attractive to bird life because of the very large surface area such plantations usually have. A wood commends itself to birds, not by having particular types of trees so much as by having a particular physical structure. The important factors are height of main growth, height and density of secondary growth, whether the canopy is open or closed, whether there are rides and other spaces throughout the wood, and so on. An open canopy is one which, when the trees are in full leaf in summer, admits light in large quantities through to the ground and the secondary growth. A closed canopy is one in which the light is shut off from the ground. Generally speaking, the more open a wood and the more exposed, and consequently better developed and more diversified, the secondary growth, the greater the bird population. Many sorts of warblers will sing from song posts on the high trees and dodge down through the open canopy to visit their nests in the secondary growth; and the greater the plant variety, the greater the available food, both plant and insect.

Perhaps the highest bird populations of all are found in areas where man has interfered with and controlled the distribution and numbers of trees (see p. 79). The numbers of breeding birds are always high in an area where there is great variety of cover and choice; and in a garden in which there are tall trees, shrubs, bushes, undergrowth, nestboxes, eaves, water, and so on, there are likely to be many nest sites, so that birds may breed in the garden but forage outside. This island effect has produced some very high populations in gardens and is now being used to good effect in modern forestry, where "islands" of old deciduous trees are left in the midst of the new plantations.

Fringing woods, parks, coppices, orchards, and gardens are artificial and man-made things, and by no means constitute the environment to which most of our British birds are naturally adapted. We find these habitats have been occupied in Britain by such birds as are fairly plastic in their selection of where to live. Blackbirds and Robins were probably originally birds of the primeval deciduous woodland, and were not so specialised, physically or psychologically, that they could not adapt themselves to the new circumstances. Bullfinches, Hawfinches and Greenfinches came to the park and garden land from scrub, Pied Wagtails probably from water meadows.

Before we leave the subject of trees, it is important to note that they form important breeding sites for birds which otherwise live in open country. This is perhaps not quite so noticeable in Britain as in parts of the great central Continental grasslands, where islands and clumps of trees are full of birds which breed there, but which spread right out through the plainland when they are foraging. You cannot have Rooks feeding in your open pastures and arable land unless there are woods for them to roost in in winter and spinneys and coppices for their summer rookeries. The only resident Rook in Shetland today is a bird that was blown there and was tamed by a crofter. There is plenty for this Rook to eat, but nowhere for it to breed. The most northerly rookery in Britain is in the most northerly wood, a clump of trees in a sheltered place in Orkney.

Rook

Birds tend to feed on agricultural land rather than to breed. But in Britain such land is bounded and divided by hedges which are so abundant (perhaps as much as a mile of hedge to every forty acres, although this figure is being much reduced in eastern and southern England) as to have a very marked effect on the bird population. Every spring the breeding places provided by these hedges are exploited to the full by buntings, finches, Dunnocks, Robins, Wrens, Blackbirds, and thrushes. A few birds like Skylarks find nest sites on the ground in the fields, but it is the hedges which keep the population as high as that of coniferous woodland and almost three times as high as that of moorland, heath and rough pasture.

There are few birds on heaths and moorland, though one or two of them are specialised and well adapted to their place in nature. Only the Skylark, Meadow Pipit, Red Grouse and, in parts of Scotland, the Oystercatcher, can muster an average population of over three birds to 100 acres—compare this with farmland. We have to go down to a density of three birds to 1,000 acres to bring in such birds as the Twite, Linnet, Reed Bunting, Tree Pipit, Wheatear, Hedge Sparrow, Ringed Plover, Golden Plover, Lapwing, Curlew, and Partridge.

Some of the birds which breed on the moors, like Curlews, Dunlin, Golden Plover, have, as their winter places, the wild shores and mudflats of the coast. Some of these are only just hanging on to their breeding grounds; today Dotterels nest only on a few remote uplands, and Whimbrel only in certain rather secret sites on northern isles. All these moorland birds have to nest on the ground or in such low bushes and whins as they may find. Even moorland birds of prey such as Merlins have to nest on the ground.

Dipper

On the very high mountain tops, moorland and heath give way to an Alpine vegetation with low creeping willows and saxifrages. Here the corner or crevice in the rock becomes the nesting site, and its inhabitants the Ptarmigan, Ring Ouzel or Snow Bunting; sometimes the Twite. These are probably birds which have had to retreat up the mountains since the world got warm after the last ice age. The Ptarmigan and Snow Bunting have been almost driven out of Britain altogether, for they only hang on on the tops of the higher Scottish peaks.

You will not find Dippers or Grey Wagtails away from streams. The Dipper is adapted in a quite remarkable way for a life in swift water. The tilt of its back against the running stream enables its feet to get a purchase on the bottom and control the direction of its movements as it moves underwater in its search for the larvae of water creatures. Both the Dipper and the Grey Wagtail nest in crevices and holes, in banks or under bridges, never far from their water.

One could prepare innumerable lists of birds and the habitats to which they are restricted. One can assign Kingfishers and Grey Wagtails to rivers and streams, Coots and Red-throated Divers to ponds and small lakes, many species of duck to larger lakes, Herons and Bitterns and Bearded Tits to marshes and fens. On the coast a salt marsh, sand dune, beach, and cliff have each their own distinctive fauna. The House Sparrow is as much a symbiote of man as the Jackal is of the Tiger.

But before we leave the subject of the habitat, we must suggest that its study in detail means much attention to the psychology and behaviour of birds. The very interesting work of Lack and others on the reasons why birds are disposed to select one habitat rather than another has already made it clear that idiosyncrasies of mind as well as of structure play a great part in determining where a bird is to live.

Perhaps the most interesting thing that becomes clear is that within one gross habitat like woodland there is a sort of share-out between the birds. Preferences for things like the height of song posts, open or closed woodland, type of nest, site, and material and so on appear to be just as (if not more) important in determining a bird's choice of world as its food preferences or tolerance of climate. As our knowledge of these factors increases, we will surely find that these psychological barriers and preferences, which birds have, help ensure that the gross habitats can be divided up in such a way that more varied species can be supported than would otherwise have been the case.

Wren

6 The numbers of birds

"Among the many singularities attending those amusing birds, the swifts, I am now confirmed in the opinion that we have every year the same number of pairs invariably; at least the result of my inquiry has been exactly the same for a long time past. The swallows and martins are so numerous, and so widely distributed over the village, that it is hardly possible to recount them; while swifts, though they do not build in the church, yet so frequently haunt it, and play and rendezvous round it, that they are easily enumerated."

GILBERT WHITE, 13 May 1778

James Fisher estimated roughly that there were about a hundred thousand million birds in the world in the 1940s; that is, there are probably a hundred thousand million rather than a million million or ten thousand million. Two attempts have so far been made to estimate the total, month of May, land-bird breeding population of Britain. The first was that of E. M. Nicholson who, in 1932, put the figure for England and Wales at about 80 millions. James Fisher's own preliminary figure for Great Britain, excluding Ireland was about 100 millions (1939). The table which follows shows a detailed revised assessment of the figures he arrived at on the latest information available to him in 1946. This puts the population at about 120 millions.

Since there are large areas of moorland in Scotland (omitted by Nicholson) and the bird population is consequently low, it can be seen that the two earlier estimates agree closely; certainly Nicholson must have the credit for being the first to give an estimate for the number of land birds in Britain. Both estimates are based, as can be seen from the table, on sample counts of breeding birds taken in the more important habitats. Much of the census work has been done by the British Trust for Ornithology, since 1961 on a greatly expanded basis. In that year the Nature Conservancy (of which Max Nicholson was the first Director General) commissioned a census of Britain's commoner birds, to be set up on a reliable statistical foundation and to be repeated each year. Its purpose is to provide a "population index" for the commoner species to make sure that no untoward changes were occurring, unnoticed, due

Habitat	No. of thousand acres 1939	Birds per 10 acres (approx.)	Total No. of birds (approx.)
Cereals	6,391		12,782,000
Roots, "cabbage", lucerne, etc.	2,161		4,322,000
Clover, etc., and planted grasses	3,361	20	6,722,000
Other crops	267		534,000
Bare fallow	370		740,000
Permanent grass	17,411		34,822,000
Rough grazing	16,081		11,256,700
Deer forest	3,430	7	2,401,000
Other ungrazed moorland	230		161,000
Parks, golf-links, etc.	33	100	330,000
Orchards	253	300	7,590,000
Small fruit	58	300	1,740,000
Hops	19	100	190,000
Gardens, allotments, etc.	313	300	9,390,000
Woodland:			
Coniferous	672	20	1,344,000
Deciduous	443	40	1,772,000
Mixed	301		1,505,000
Fringing and coppice	734	50	3,670,000
Waste, scrubland and swamp	939		4,695,000
Built-up*	2,740		13,700,000
Inland water	593	10	593,000
TOTAL	56,800		120,259,700

* Approximate

to changes in pesticide usage or agricultural practice. Each breeding season, hundreds of birdwatchers visit their census plots (of between 50 and 400 acres) from 8 to 15 or so times, and record on large-scale maps the position of territory-holding birds. At the end of the season, these sightings can be brought together on a tracing and the position and numbers of the various territories outlined. The results are compared with the previous years, and population changes over the whole country can be determined with remarkable accuracy. As a by-product, new density figures are available, and I have combined data from the Common Birds Census (CBC colloquially) reports published each year in

27   Bagley Wood—CBC maps showing how individual visit sheets (top map) can be combined to produce species maps at the end of the season. The species shown are Whitethroat and Robin, and the technique of marking the cluster of records to indicate territories can be seen

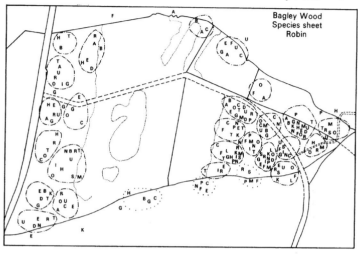

Bagley Wood
Species sheet
Robin

Bird Study with land-usage and acreage details from Ministry of Agriculture tables from "Farming in Britain Today" by J. G. S. and Frances Donaldson and Derek Barber, to produce estimates of the situation in 1972.

Habitat	Millions of Acres (approx.)	Birds per 10 acres (approx.)	Approx. total birds (millions)
Farmland:			
(Arable, mixed inc. fruit)	29	24	69.6
Rough Land:			
(inc. moorland and hill grazing)	17	5	8.5
Scrub etc.	2	70	14.0
Forest and Woodland	4	57	22.8
Urban areas	3	60	18.0
Water	1	10	1.0
TOTAL	56		133.9

It will be seen that I have used rather broader habitat divisions than the previous estimates because these fit in better with the nature of the sample areas surveyed under the CBC. Nevertheless, the agreement with Nicholson and Fisher is surprisingly close, and the small increase that it shows is comforting in days when we feel (subjectively) that our birds may not be faring too well. Here the CBC index figures come to our aid again, because they show how many of our birds are faring with the passing years, and with natural changes, such as climate, or with unnatural or man-made habitat change. While many of these may be described aesthetically as desecration, they may not be as harmful to many of the more adaptable of our species as we might suppose. The more specialised birds (for example those of our rapidly dwindling wetlands) do, regrettably, suffer more severely. But in numerical terms, if they are excluded, more practically generalised species move in to the modified habitat—perhaps in greater numbers, so that at this stage we should be more concerned to maintain a rich diversity of species (which means a similar diversity of habitat, and retaining, by protection, workable-sized areas of specialist zones) than worrying about total numbers.

In general terms, the mild winters of the late sixties and early seventies have helped population of some of our smaller birds greatly. These are the ones that suffer in severe frosts and snow—like the Wren, or the Goldcrest in Figure 28, which gives some striking examples of the ups and downs of the CBC population index.

Though there are 473 species in the British list, only about 100 of them contribute significantly to this total and the great bulk of it (about 75 per cent) is made up by 40-odd species. James Fisher thought that the commonest birds in England and Wales were the Chaffinch and Blackbird, of which there were about 10 millions each. He reckoned that there were about 7 million Starlings and the same number of Robins. Based on CBC results, it is possible to derive some new "league tables", but these, it must be remembered, can only be the roughest of estimates. The latest figures (1972) provide a basis for the following estimates of the number of individuals of various species present at the start of the breeding season:

Blackbird	15 million
Wren	10½ million
Robin	10 million
Chaffinch	8 million

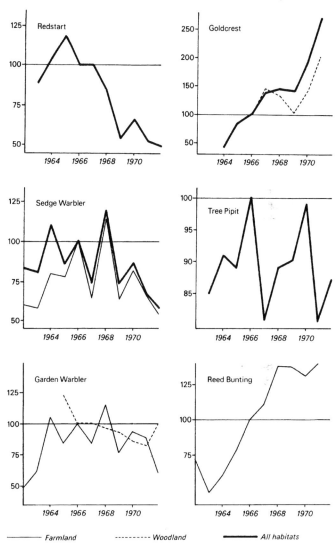

28 Nesting populations of 6 species in Britain—graph compiled from CBC reports over a ten-year period (1966 = 100)

Dunnock	8 million
Willow Warbler*	6 million
Song Thrush	6 million
Blue Tit	6 million
Skylark	5 million
Blackcap*	2½ million
Whitethroat*	2 million
Chiffchaff*	2 million

* summer migrants

From this top dozen there are two deliberate omissions: the Starling and the House Sparrow. Both are numerous and widespread, the Starling in many habitats, the House Sparrow especially in association with man, in towns and on farms alike. Regrettably neither lends itself to census techniques, but my personal feeling is that by now our House Sparrow population may have ousted the Blackbird from "number one" spot, perhaps approaching the 20 million mark. The Starling would, I suspect, feature in the middle orders. A glance at Figure 29 shows that up to 1968, the Whitethroat would probably have been "top migrant"—clearly it will be many years before it reaches that position again.

These are the birds with large populations—so large that they can be arrived at with some degree of accuracy from taking counts of random samples of the countryside. With birds of smaller populations the

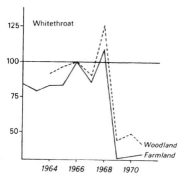

29   Whitethroat nesting population in Britain for a ten-year period showing the 1969 "crash"—from CBC reports

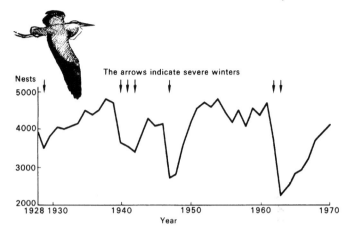

Nests

The arrows indicate severe winters

30 Heron—graph showing numbers of nests in Britain from 1928 and the effect of severe winters

numbers are so few that, unless one takes a very large sample, the total population cannot be accurately estimated. Some other method must be invented for counting them. As far as land birds go, it has only been possible to count a few populations.

In Britain at the time of the war there were about 150,000 Black-headed Gulls, a figure which rose to about 200,000 in 1958 and must be considerably higher by now. These can be included amongst land birds as so many of the colonies are many miles away from the sea. There are some 8,000 Herons—and oddly, this seems to be a ceiling figure (Figure 30). Severe winters cause the numbers to drop dramatically, but each subsequent rise seems to level out at between 8,000 and 9,000 birds. A 1931 census of Great Crested Grebes showed about 2,800 adults—a repeat, in 1965, showed about 4,500. Possibly this increase arises in part from protective measures, but more likely it is based on the tremendous post-war boom in gravel extraction (for rebuilding purposes) and consequent new stretches of suitable water.

It has been possible to count these birds because of the existence of co-operative observers in all parts of the country (chiefly members of the British Trust for Ornithology) and because of the particular habits of the birds themselves. Herons and Black-headed Gulls nest in very obvious and sometimes spectacular colonies, those of the latter in one case reaching on occasion nearly 50,000 pairs (Ravenglass, in Cumberland, which had about two-thirds of England and Wales' population). In many areas where there are more than a certain number of observers, these colonies can all be found and counted, and this has been done in these particular cases, though the Black-headed Gull figures have had to receive certain mathematical treatment before they could be finally estimated.

The only other land birds whose population can be arrived at with any accuracy have up to now been those which are extremely rare. Quite often a pair or two of birds are the only ones known to be breeding in the country and one can arrive at such obvious figures as two for the population of the Black-necked Grebe in England in 1939, about twenty for its population in Scotland, and under 250 for its population in Ireland. Kites are to-day breeding only in Wales, and where James Fisher could report only four nests, continued perseverance and protection by the RSPB and the Nature Conservancy has resulted in about 20 nests each year at the moment. Despite this protection, even in 1972 eggs were lost to collectors—a sad reflection on our times. How much more beautiful and useful the birds than their sterilised eggs—contents blown out, fertile or not, kept as pieces of coloured chalk in secrecy in a locked cabinet. What is the point?

Some stories are sad ones—in 1946 James Fisher would not have considered the Red-backed Shrike rare—pairs could be found on any chalk downland with hawthorn bushes to provide the spikes for the "butcher birds" larder to be hung on. Now we have perhaps fifty pairs, and the number falls still. The reasons are difficult to uncover, for although persecuted by egg collectors, plenty of habitat remains, and the theory that the appropriate insect food is in short supply is hard to substantiate.

However, all is not gloomy by any means. In 1946 the two great success stories had not begun—the return of the Avocet and the Osprey as British breeding birds. These stories are a triumph for the purposes of protection—in the one case requiring engineering works to protect the habitat, in the other electronic devices to protect the nest from those anxious to obtain eggs.

One rare species has a population which is pretty accurately known. This is the St Kilda Wren. In 1931 the Oxford expedition recorded a

Osprey

total of sixty-eight pairs of this interesting little bird, and the visit paid by Dr Julian Huxley, E. M. Nicholson, and James Fisher in 1939 proved that no great change had taken place in its numbers.

In 1957, Ken Williamson estimated 120 pairs following a dawn census of singing males, and in 1969, a similar census by David Musson and Jim Flegg showed about 140 pairs. It might be noted here that the proper estimation of bird population needs a lot of practice; it is very easy to over-estimate the numbers of birds, either directly or by implication.

So much for the population of land birds. The numbers of sea birds are extremely large on a worldwide basis, but not by any means as large in Britain as those of inland birds. Nobody knows much about the precise numbers of these birds (except for certain rare species and others which have been the subject of specially intensive study). A rare bird, Leach's Fork-tailed Petrel, breeds on only three or four islands off our coast and probably cannot muster more than 4,000 individuals, though the true population is difficult to assess, except at night, when the number can be roughly estimated by the sounds made by the incubating birds in their burrows.

In 1969, our coastal seabird colonies were censused by the Seabird Group in "Operation Seafarer" (which owed much in its genesis to James Fisher). The results, published in *The Seabirds of Britain and Ireland* by Stanley Cramp, W. R. P. Bourne and David Saunders, show the numbers of three species at the breeding season to be about the million mark—Guillemot, Kittiwake and Puffin—and that the total seabird population must be well in excess of six million individuals.

By means of a rather complicated mathematical dodge the popula tion (in 1939) of breeding Fulmars in England and Wales was put at about 480 and in the whole of Great Britain and Ireland at about 120,000. This figure is dwarfed by the continued spread recorded in Operation Seafarer, when over 300,000 *pairs* were recorded. There is only one bird whose world numbers are not small and which are anything like accurately known. In 1939 a group of ornithologists, including H. G. Vevers and James Fisher, visited nineteen out of the twenty-two gannet colonies in the world, and were able to make counts of all but 2 per cent of the birds. They believed that there were about 165,600 gannets breeding in the world in 1939, of which 109,100 bred in Great Britain and Ireland, and about 11,800 in England and Wales. All but about eight of the latter bred on the colony of Grassholm in Wales. The eight come from what appeared then to be a new colony trying to establish itself on the Yorkshire coast. This now forms part of the RSPB reserve at Bempton. Operation Seafarer has shown a new colony in Scotland, and large increases in those already known, to the extent that the 1939 *world* population was almost doubled by the 1969 British and Irish figures (138,000 *pairs*).

Besides the gannet there are relatively few birds with known populations world wide, and most of these are documented in the IUCN *Red Data Book* of threatened species. Many must be on the verge of extinction, like the Japanese Crested Ibis (12 in 1965), Mauritius Kestrel (10 pairs) and Whooping Crane (33 in 1959, 50 in 1968). Others, though, have been brought back from the verge. In 1946, Fisher listed only 50 remaining Hawaiian Geese (Ne ne) but an intensive and skilfully managed rearing programme using selected parents, at the Wildfowl Trust at Slimbridge, and later in other collections, has raised the numbers sufficiently to allow some hundreds of captive-reared young to be released again to the wild on a reserve specially purchased for them on the island of Maui.

In order to establish with reasonable accuracy world population levels, one or more of the following conditions must be met,

(a) That they are social sea birds with a limited number of colonies and a definite breeding season, so that expeditions can count the number of nests at each colony.

(b) That they are social birds with a very restricted geographical distribution, and a limited number of colonies irrespective of whether they are sea birds.

(c) That they are large, spectacular, and easily recognisable birds (like the Trumpeter Swan and Californian Condor) which are nearly extinct or which have a relatively small breeding area, in which the territory or hunting ground of each individual pair can be discovered.

(d) That they are species of birds restricted to small islands.

THE COMMONEST BIRD IN THE WORLD

Darwin once suggested that the Fulmar was the commonest bird in the world, but judging from its world breeding distribution, James Fisher was inclined to disagree with him. Modern research has shown that this distribution is far more restricted than was previously supposed. Certainly the number of Fulmars must not be judged from the outpost population in the British Isles. In Iceland and the Faeroes, Jan Mayen, Bear Island, and Spitsbergen there are millions of Fulmars. There may even be tens of millions, but I should very much doubt whether there are hundreds of millions. Darwin may have got his idea from the abundance of the Fulmar at sea in the North Atlantic. Certainly it is one of the commonest birds in that part of the ocean, but James Fisher suggested that the Little Auk, another arctic breeding species, runs the Fulmar fairly close if it does not actually beat it. Little Auks breed in fewer places than Fulmars, but in enormous, closely packed colonies on such arctic islands as they inhabit.

It is very difficult to assess large numbers. It needs a great mental effort to think in the necessary terms, and comparative numbers become as hard to imagine as actual ones. All the same, the most abundant bird in the world is certainly likely to be a sea bird, and probably, Fisher thought, Wilson's Petrel. Curiously enough this creature has only been observed in Britain about a couple of dozen times. Its breeding haunts are the Antarctic and some of the surrounding islands, and there its numbers, Dr B. B. Roberts says, are quite fantastic. In the off season, Wilson's Petrel moves into the oceans, and huge flocks pass up the Atlantic, Pacific and Indian Oceans, and even into the Red Sea. As they get north they tend to congregate on the west side of the Atlantic, reaching their peak in numbers in our northern autumn in the neighbourhood of Long Island. There are probably one or two other seabirds which may run Wilson's Petrel fairly close: these are the Cormorants of the Cape and of the Guano Islands of Peru.

It is possible to guess at the most abundant land bird. It may lie between the Starling and the House Sparrow, but the enormous numbers of some of the weavers (*Quelea*) in Africa should not be overlooked. Their flocks are so huge (sometimes a million or more

House Sparrow

strong) and so mobile that accurate assessment is unlikely. Both House Sparrow and Starling have origins in the European continent, where they are very widespread, and have been transferred by human agency to other continents. In North America both Starlings and House Sparrows are pushing rapidly westwards, where they are in many regions becoming dominant birds. The House Sparrow is getting to other parts of the world as well. It has reached Uruguay and arrived at the Falkland Islands in 1919. It was introduced into New Zealand in 1862, and into South Africa at the end of the nineteenth century. Though the House Sparrow has now a wider distribution in the world, the Starling has probably the higher population, since it is not so restricted in its habitat.

The most widely spread land bird in the world is pretty certainly the Barn Owl. Its races extend to every part of the world except the polar regions and New Zealand and certain of the Pacific Islands although it has penetrated as far as Fiji.

METHODS OF COUNTING

There are two main methods of counting birds, the area census and the breeding count of an individual species. Both methods are valuable for finding actual populations and for comparing the situation in different years. Again we must emphasise that our general idea of land-bird breeding population has been arrived at from the first method, which is perhaps the most important. The general technique is to take a sample of one of the major habitats, to find the area of this sample, and to find the total number of breeding pairs of all species in it. In the long run the only absolutely accurate method is to find every nest, a method which needs much skill and practice and which, in honesty, is never likely to be achieved over a suitably-sized area.

Most of the intensive work on the bird life of farmland, woodland, moorland, and so on in Britain, is being done by members of the BTO, using the CBC techniques described earlier, and we have a picture of the relative abundance of the different kinds of birds which is a good deal more accurate than our idea of their actual total numbers. Of course, relative numbers are of great biological importance, so it is not surprising that much work and thought has been given to their discovery. The simplest method of all is to walk through a selected part of the habitat and make a tally-list of the species and the numbers of each seen or heard.

Blackbird

COUNTING THE INDIVIDUAL BIRDS

We have already suggested that most of the counts of individual species of birds have been successful when the birds have been social breeders. This particularly applies to sea birds. The main task of the Gannet-counting teams during their world survey in 1939 was not so much the actual counting as getting to the colonies. In this particular case the colonies are on the remotest possible stacks and rocks (see

Plate 2), like Eldey, 20 miles from the nearest fishing village out in the Atlantic off the coast of Iceland. St Kilda is 45 miles west of the outer Hebrides. Sula Sgeir is about the same distance northwest of Cape Wrath. Sule Sack is about 35 miles from Cape Wrath and the same distance from Orkney. The Hermaness (Muckle Flugga) colony is the northernmost tip of the British Isles. Once the colonies have been reached, there are certain further technical difficulties, such as those of landing and climbing, but the task of counting each occupied nest is comparatively simple, though it imposes a certain strain on the eyes.

Most people who visit a place where birds are breeding carry away some impression of population. Usually this is interpreted in their notebooks as few, common, abundant, and so on. It needs little extra effort to make these notes entries less vague, and one's rapid impression can quite easily be translated into terms of under ten, under 100, under 1,000 (and so on) breeding pairs or breeding birds. Of course, notes on actual numbers are far more useful than notes of the above kind, but those general figures are far better than no figures at all, and can often be interpreted to a good deal of advantage by somebody who is working up the particular species later. Our knowledge of the Fulmar's increase in population (as opposed to its simple geographical spread) is largely due to interpretation of observations in terms of under ten, under 100, and so on.

We have already noted that the numbers of birds fluctuate, often in an orderly and remarkable way. In Chapter 4 we have discussed some of these fluctuations, especially those of species like the Crossbill and Pallas's Sand-grouse, and Figure 28 has shown us others. We should not forget the nature and size of these fluctuations when making comparisons. Nevertheless, there are many trends in the numbers of British birds which we cannot yet ascribe to periodic fluctuation. Some species are tending to increase and spread, others to decrease and disappear from their previous haunts.

POPULATION SPREAD

Perhaps the most spectacular spread in recent years has been that of the Fulmar. In the year 1697 this bird was known to breed on the islands of St Kilda, and though there are one or two early records of colonies elsewhere, none of these can be really substantiated. In 1878, however, a colony established itself in Shetland on the island of Foula, and since that date the number of known colonies has increased to such an extent that in 1939, 208 different stations were known at which birds were definitely breeding, and 61 others at which birds were present in the breeding season but had not yet been proved to breed. The spread took place, first, to the more oceanic headlands, followed by the colonisation of the intermediate cliffs, and a gradual pushing up the firths and locks. Down the coast of Britain Fulmars reached the limit of high cliffs in Yorkshire in 1922; by 1940 they were prospecting the lower cliffs of Norfolk, and in the breeding season of 1944 were seen at cliffs in Kent. In 1944, too, they first bred in Cornwall and on Lundy Island off the Devon coast. At about this time, they were seen, in summer, at cliffs in Dorset and the Isle of Wight, and were present every year in Pembrokeshire; they had pushed down the North Channel into the Irish Sea, and bred on the Isle of Man and in County Dublin. They were present in the breeding season at Holyhead, and bred in 1945 on Great Orme's Head. In Cumberland they first bred in 1941, at St Bee's Head. They have pushed up the Inner Hebrides from the south. The series of maps in Figure 31 shows roughly the way their spread has taken place.

Most of the colonies, except for some in the very north of Britain, are still small; their numbers have increased much more rapidly than the total population of Fulmars, which has trebled in its sixty years of imperialism. This trebling is, all the same, remarkable, since the Fulmar usually lays but one egg a year and rarely replaces it if it is broken or lost, and it may only breed once every three years. At certain times the increase has been at over seven per cent compound interest, and through most of the time at over three per cent per annum.

The causes of this remarkable change in numbers and distribution is not properly known. It was recently suspected that the increase might have been due to an overflow from St Kilda, since at roughly the time it started the St Kildans began to receive supplies of food from the mainland and were supposed to have discontinued their famous fowling habits. But an exhaustive survey of the literature about St Kilda, which is very complete, shows that, until 1910 at least, the St Kildans went on taking birds as much as ever. Moreover, it is becoming clear that the Fulmar started spreading many years before the first new colony was established in Britain. The spread dates, from about 1816, when the bird began to increase greatly in Iceland, and some time between then and 1839 it arrived at the Faeroes, north-west of the Shetlands. In the

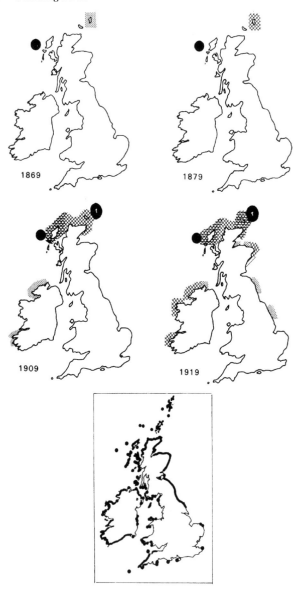

1869

1879

1909

1919

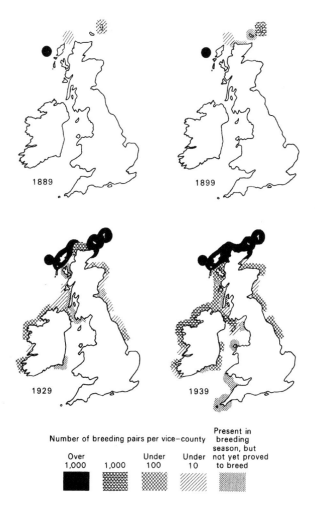

31 Fulmars—maps to show the spread of breeding in recent years—inset in-dicates breeding range (not numbers) in 1972 (from John Parlow's *Breeding Birds of Britain and Ireland*)

Faeroes it spread so much that the inhabitants were taking and eating 100,000 young a year, which was very much more than the total number of young produced in one year in Britain. In 1936 this taking of young Fulmars was stopped, since an outbreak of psittacosis was traced directly to the birds. Writing on the situation in 1966, James Fisher expressed his conviction that the prime cause of this expansion was the increasing provision of fish offal from modern fishing fleets, with the majority of the catch gutted and frozen or salted at sea.

There have been many other recent and remarkable spreads of species of birds in Britain. The Little Owl, which was introduced into Britain (in Northamptonshire) in 1889 and in Kent in 1896, has now spread to every corner of our country, as far north as Lancashire and Yorkshire. It reached Norfolk in 1912, Land's End in 1923, Pembrokeshire by 1920 and the Pennines and North Wales by 1930. The reasons for its spread are relatively clear. It was a Continental bird from mid-Europe, separated from our country by wide natural barriers. Once these had been overcome by importation the bird found that the place it occupied in nature in its native haunts was not occupied strongly by any bird in Britain. Without competition, its increase was assured.

There have been some signs of recent invasion from the Continent by one or two species which seem to be generally extending their range in the west. The Avocet and the Osprey have already been mentioned, but the Black Redstart seems to be arriving in increasing numbers year by year, and every season brings new records of its breeding in southern counties. It has bred in the heart of London since 1940. The Stock Dove has invaded Ireland presumably from England and Wales: since 1875, when it was first noted, it has increased and spread, till to-day it is breeding in nearly every Irish county. At the same time this bird pushed into the north of Britain and invaded Scotland. To-day it breeds in the west as far as Argyll and on the east right up to Sutherland.

Two wetland warblers are new arrivals, Savis Warblers arriving in Kent in the 1960s, and breeding there and in other south-eastern counties (but in low numbers still), and more recently Cettis Warbler. This latter is a non-migratory warbler, and thus its tenure in Britain will depend much on future winters' severity. In the last three years it has cropped up, with its noisy song, in a number of south-eastern localities and perhaps a couple of dozen pairs are now breeding.

The success story to cap all others concerns the Collared Dove: from first sightings in Britain (greeted with great scepticism) in 1952, to first

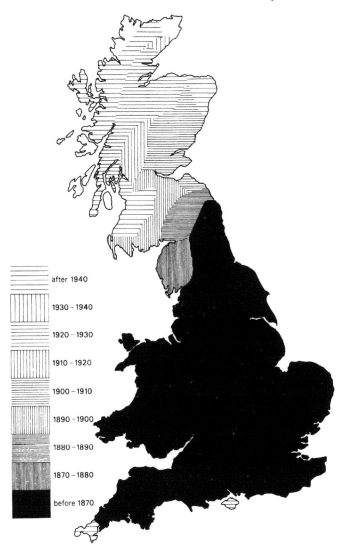

after 1940

1930 – 1940

1920 – 1930

1910 – 1920

1900 – 1910

1890 – 1900

1880 – 1890

1870 – 1880

before 1870

32   Greater Spotted Woodpecker—spread of this species during the last hundred years

breeding in 1955, the Collared Dove has since conquered all of Britain and Ireland, breeding even on the west coast islands of Ireland, occurring regularly on remote St Kilda and now breeding in Iceland! All in the space of 20 years. Quite what lies behind this westward surge, is uncertain. Presumably once in Britain, it found no competitor strong enough to deny it its chosen niche, but what caused the original eruption from the eastern end of the Mediterranean, colonising the whole of Europe? Perhaps a sudden genetic change—but then how did this spread so rapidly through the population?

The Greater Spotted Woodpecker is another bird which is pushing its range forward. In recent years it has become very much more common in the north of Scotland. In 1939 it was definitely recorded from a wood in Sutherland not far from the Caithness border, and lately it has been spreading through Argyll (see Fig. 32).

Many other examples of increase of range might be quoted, and examples of species which are finding new habitats as well as new lands to conquer. Thus Curlews have increased and spread onto relatively low ground in England in the last twenty years. Black-headed Gulls are tending to breed more widely inland, and are to-day land birds as much as sea birds. Their numbers are increasing too as we have seen. In the north of England and Scotland, Oystercatchers have spread up the river valleys and are breeding many miles from the seashore. Reed Buntings are becoming more and more birds of scrub; occasionally, woodland and even gardens in winter (where they favour bread!), and the Siskin is now a regular attender at many garden peanut holders.

MAPPING SURVEYS

In 1968, the BTO launched possibly the biggest birdwatcher participation exercise ever, when a five-year programme to visit all the 3,850 ten-kilometre squares of the National Grid and to record their breeding birds was started. Amazingly, the target of complete coverage was achieved by 1972, and so far as can be judged the standard of observations was very high. Each "square" surveyed meant preliminary consultation of large-scale maps to find the best-looking sites and to ensure that all habitat types were covered, and the planning of an effective survey-route. During the breeding seasons, visits were made in different months and at different times of day. The techniques evolved are now becoming standard practice for birdwatchers visiting

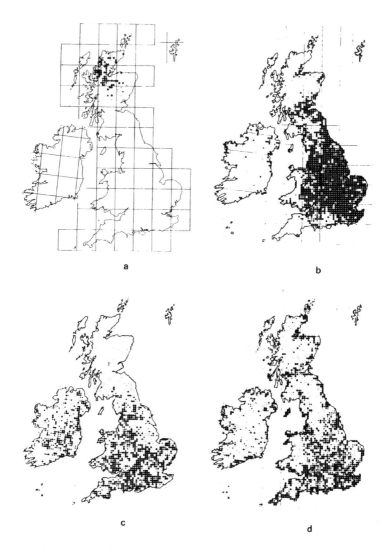

33   Provisional distribution maps from the BTO atlas of breeding birds: (a)
Redwing; (b) Tree Sparrow; (c) Kingfisher; (d) Collared Dove

new areas, and many observers have been startled to find just how prolific were parts of their own parishes that they had not bothered to visit. Figure 33 shows some interim examples of the maps now being produced. Not only were they the basis of some most enjoyable birding, but the final product will be of great biological interest. For example—what are the features which produce a northwesterly distribution for the Pied Flycatcher, a southeasterly one for the Nightingale? Why is the Corn Bunting so erratically distributed, and what causes the Yellow Wagtail to be a water-meadow bird in one county and a potato-field bird in another? Certainly in this sense, the maps and all the possible correlations with geological features, vegetation, climate and land use, will pose more questions than they answer. From the conservation point of view, the breeding situation of Britain's birds will be "frozen", and this frozen picture will serve as a yardstick against which future changes can be assessed. Already the scanty remnants of the once-widespread distributions of Woodlark, Nightjar and Cirl Bunting have been revealed, showing the areas in which these species are in need of protective measures, and where such measures can sensibly be considered.

POPULATION DECREASE

Some of the causes of population spread are hard to define. So are some of the causes of decrease. It is difficult to imagine why it is that Wrynecks have become so increasingly rare in the last few years—though it has been plausibly suggested that this is due to the cutting down of old gnarled orchards and the planting of standard apple trees; that is, to the destruction of many of their favourite nesting sites. It is difficult too to explain why Rock Doves are now no longer breeding in England or Wales or on the east coast of Scotland south of the Firth of Forth.

Many of the decreases of British birds are, however, attributable to human agency. These causes, it should be stated, are in the long run often deep-seated and probably out of the control of the conservationists. It was the draining of the fens and not the absence of bird protection laws which lost us Cranes and Spoonbills as breeding species. The cutting up and working of plains-land by man drove out the Great Bustard as a breeding species by 1833; and is now making the Stone Curlew a rarish bird, breeding on such barren downs as it can still find. It is highly likely that one of the most spectacular recent decreases, that of the Corncrake, is due largely to human agency; the results of a British Trust for Ornithology investigation (in which about 2,000 observers took part) showed that the development of the modern mowing-machine, together with the earlier date of grass-cutting in more intensive forage cultivation (at least three weeks in advance compared with 50 years ago), has been very closely linked with the Corncrake's disappearance. An enormous number of Corncrakes and their young must have been killed by the mowing, and the changing patterns of agriculture must discourage many others.

The main task of those concerned with bird protection seems to be threefold. First, their job is to ensure the continuance of certain areas in their present natural state so as to preserve a sample of bird life as little affected by man as possible. Secondly, their task should be to preserve the different species of birds because of their population, interest, and biological importance as much as because of their rarity. Thirdly, they have to keep the collector away, particularly from such species as are on the verge of extinction. There is no need to fear anyone in saying that the extinction of a species should never be ascribed to collectors in the long run, though they may often be the last straw. The fact that the remaining pairs of Kites to-day are jealously guarded from collectors by a private army of enthusiasts certainly does not mean that collectors alone have brought matters to their present state. The trouble is that when for various reasons a bird becomes rare, its importance and the importance of its eggs from the collector's point of view become correspondingly greater, so that its end is hastened. If collectors were finally prevented or educated away from breaking the law, as we must hope they will be, it is possible that a remnant of many species now nearly extinct would be saved. But it is equally probable that this remnant would seldom become anything more than a remnant, and that while it would remain very interesting from a human viewpoint, it would never play a serious biological part in our bird fauna.

BIRD PROTECTION

It is possibly apt to make some remarks in this chapter about the organised protection of birds. First we must realise that in this country we have inherited a wonderful bird fauna. It is rich in species, rare and common, of simple and curious habits, harmful and beneficial, dull and exciting. When we consider the preservation and conservation of this heritage of birds, we must have several attitudes in mind, and we must be careful that our special fancy for any one of these attitudes does not lead us to neglect the weight of the others. Our attitudes must be biological, scientific, economic and aesthetic. We must only allow ourselves to be sentimental in so much as it allows us to satisfy the other more important attitudes. We must see ourselves first as animals living in this country in an ecological relationship with the wild creatures around us.

If a 500 years old tradition encourages men from the Isle of Lewis to make a hazardous boat journey every year to take young Gannets on Sula Sgeir, we must try to arrange that they are able to pursue this, to them, important and economic custom. We must not say this killing must stop; neither must we say that we can leave the future of these Gannets to chance. Our solution is to apply what knowledge we have of the vital statistics of the Gannet, and to ration the numbers taken for the benefit of both man and bird. The men of the Faeroes, who take Gannets every year from the Holm of Myggenaes, farm their birds as carefully as if they were chickens.

Most of the common small birds of the countryside are probably beneficial in the long term, and if the amounts of good and harm done to human agriculture and enterprise by birds as a whole are totted up, they come out about level, or better. The harmful birds comprise relatively few species, though these do altogether as much harm as the many beneficial species do good.

The biological value of the Rook is very imperfectly understood. Quite serious controversies still rage about whether it is ultimately beneficial or injurious. Certainly the most injurious bird in Britain is the Woodpigeon, although the House Sparrow may by now be catching up, with its haul of ripening cereals. The Woodpigeon probably does much more harm than the most beneficial bird does good.

The aesthetic argument is the stimulus behind practically all the bird protection that has been organised by law and active deed in this country. Among the many facets of this argument, a new one is becoming important—that birds should be preserved, not only because of their rarity, curiosity, wildness, or romance, but also because of their biological interest and because of their involvement in the complex inter-relationships of the web of life. This new argument contradicts some of the old ones, for many of the predatory birds disliked by sportsmen and sentimentalists are of primary interest to biologists.

It is worth quoting James Fisher's views (in 1946) in full: "In my opinion those who have framed the policies of bird protection societies and who have drafted the standing orders to watchers of sanctuaries, have not fully asked themselves the whys and wherefores of their protection. They have not always decided what birds they are going to protect and what they are going to protect them from. The public mostly thinks of protection as being bound up with sanctuaries, and indeed most of the work of societies like the Royal Society for the Protection of Birds is concerned, when it is not enforcing the existing laws, with the management of smallish areas of land for the benefit of birds. Some of these sanctuaries are probably below the minimum size in which active management would really pay by producing the goods (whether the goods be more rare birds or better places for birds to live in). But many of them succeed in keeping mischief-makers away from birds which are particularly interesting, on the threshold of existence, or which are of some special sentimental importance.

"It should be stressed that bird protection can never be really efficient until sanctuaries can become national parks, individual problems become local problems, and local problems become national problems. We cannot really know what is to become of our birds, and from this knowledge manage them, until we treat them like any other national commodity such as sewers or the unemployed or electricity."

Clearly, since then we have made giant strides. But have we as individuals (let alone as a nation) really assessed the problems yet? Have we come to terms with the situation and accorded priorities, and above all are we ready to pay the going rate to operate a conservation policy? I feel still that in many people the emphasis is wrongly placed—the "rarity value" and "zoo appeal" of bird and reserve alike remains more important than the aesthetic needs, demanding wide, unchanged acreages or the ecological demands that the so-prevalent trends towards monocultures (or at least oligocultures) should be slowed or halted. For their part, the ecologists who are at present in some danger of being over-run by public awareness and concern—the "environmental revolution"—must continue to work to produce factual evidence in support of their theories, however difficult and time consuming the tasks. In the long run, an over-played hand can be as dangerous as lethargy.

We now have legislation on the lines propounded by James Fisher in

1946, but it is *still* confusing, and still difficult to apply. The RSPB are to be heartily congratulated at beating both the offender and the treacle tart of the law in each successful prosecution. But how much better it would be if we felt concern for our birds (and of course all wild things, plant or animal), regarded them with proper concern (for their ills may foretell ours, as does the miners' canary) and considered this complex of islands as one whole nature reserve, treated with consideration and despoiled as little as possible. We complain of the lack of a national strategy to guide and rationalise planning, to eliminate the running sores of hotch-potch development: is this lack a reflection of our disinterest, or inertia, or selfishness? We all use petrol, but none of us want an oil refinery nearby!

Great Crested Grebe

7 Disappearance and defiance

"The note of the Whitethroat, which is continually repeated, and often attended with odd gesticulations on the wing, is harsh and displeasing. These birds seem of a pugnacious disposition; for they sing with an erected crest and attitudes of rivalry and defiance; are shy and wild in breeding-time, avoiding neighbourhoods, and haunting lonely lanes and commons, nay, even the very tops of the Sussex Downs, where there are bushes and covert."

GILBERT WHITE, *2 September 1774*

The frontiers where the world of birds impinges on that of man are largely those of emotion and symbolism. The ordinary child or man with a normal interest in birds thinks first of them as singing, or building nests, or chasing each other, or flying in excited flocks—as doing, in fact, the things that are the most highly symbolic or stereotyped of all their habits.

Many of man's habits are highly symbolic, and though born no doubt for some useful function, have, in the course of time, lost much of their immediate use. Yet they are retained, since they fulfil our emotional needs. Such habits are singing, dancing, telling stories, making puns and jokes, a lot of smoking, the curious, interesting, and rather pleasant behaviour known as "manners", and much religious behaviour. Most of these habits are instinctive (a word which must be used with care) and few are intelligent—though that need be no reason for deserting them.

Birds have no intelligence to speak of—in the way in which we popularly understand the word. Their emotional, instinctive, and symbolic actions are correspondingly all the higher and more important. In many ways their very emotionalism is robot-like. Their world is so narrow that normal life presents them with a very limited number of situations—and the number of possible responses they can make is just as limited. It is not surprising, therefore, that when presented with a situation quite unknown in their normal life, birds should react by doing something which does not fit (but which is itself normally a reaction to something familiar to them). Thus a gunshot may startle birds in a wood

115

into full song; a man, suddenly intruding, may find Adelie Penguins offering him stones, nesting Stanley Cranes offering him sticks, pheasants in captivity may court their food-troughs or their keepers. If a man had a bird's mind he might quite well be given to reciting "The Boy stood on the Burning Deck" (or some other more suitable epic) at a smoking concert, but might also recite it in an air-raid or an earthquake.

Within the web of the bird's own ordinary life, its limited repertoire of actions and reactions suits it very well; and as so many of its habits of life are symbolic, like song or courtship, and for that reason curious and interesting to us, we must make some effort to examine them.

The symbolic actions of birds—those actions designed to appeal or signal to other animals in their immediate world (whether birds of their own kind, other birds, mammals, including man, or any other sorts of animals)—may involve sight, sound, or touch, but rarely smell (birds have very little sense of smell). When these actions involve some sort of posturing, in which colour, shape and adornment may be used, they are often called displays.

CONCEALMENT

One of the most interesting ways in which birds "signal" to other animals is a purely negative one—in the sense that it is the opposite of signalling. It is the habit of concealment. At different stages of their lives, many different sorts of birds have concealing colouration, which enables them to "disappear" into their immediate environment and thus hide from their enemies, or even, in the case of (say) the Tawny Owl, to blend with their surroundings so that potential prey, unawares, approaches their perch sufficiently close for the owl to "parachute" down onto it. Many British birds show concealing colouration or shape. Examples are:

Eggs—Those of the Ringed Plover, laid on stony ground, resemble coloured pebbles, and are very difficult to see (Plate 12). The eggs of the Lapwing (another ground nester) have a camouflage pattern which breaks up their outline. Many other waders and gulls have eggs with similar protective colouration.

Nests—Most land birds hide their nests, many by concealing them in thick bushes, in leafy trees, in long grass, and so on. But some go farther, and build their nests to resemble some of the special, rather than the general, features of their environment. Thus a Wren may cover its nest with moss and lichens to resemble the mossy stump in which it is placed, and the neat, mossy nest of the Chaffinch is well concealed in the fork of a tree. Some young birds in their nests complete an illusion; thus the young of the Ceylon Black-backed Pied Shrike look like a branch stump (Fig. 34).

34   Nest of the Ceylon Black-backed Pied Shrike built to resemble a mossy branch stump, with young birds seemingly part of the branch outline

Young Birds—Many fledglings, particularly those which leave the nest as soon as they are hatched, have protective colouration consisting often of a disruptive pattern which breaks up their outline. Newly hatched Oystercatchers, Ringed Plovers (Plate 12), and terns are almost invisible. Young Ringed Plovers also have a "signal" patch of white on the back of the head, which is clearly visible when the bird is upright with neck stretched. If, when frightened, it crouches, the patch is concealed.

Adults—Partridges, hen Pheasants, Woodcock, Snipe, Bitterns, Nightjars (by day), and several other birds have concealing colouration by means of which they blend with the general surroundings. This property is chiefly brought into use when the bird is on the nest. The Poor-me-one on its nest usually looks as much like a branch-stump as the young of the Shrike in theirs, and the Bittern, black-striped, neck erect, beak pointing skywards, blends into the reeds if surprised whilst hunting.

Bluff—Many weak and inoffensive birds defend themselves against possible enemies by means of bluffing. Often, people are suspicious of hissing from hollow trees or nestboxes, feeling that a snake may be inside. In nestboxes, the "snake" may be a Wryneck, stretching out its neck and hissing as hard as it can, and tits and owls (and even

Goosanders) will hiss in an intimidating fashion when their nests are disturbed.

Mimicry—Among the animal kingdom we find examples of the mimicking of distasteful or pugnacious animals (models) by harmless ones. As long as the mimics are fewer than the models, this works to their advantage, since predators, when sampling the group, are more likely to capture a model and be thus disinclined to repeat the performance. In Britain we have no good examples of birds with these characteristics, though in Madagascar we find harmless thrush-like birds mimicking pugnacious shrikes, and in the East Indies orioles mimicking Friar Birds. It is possible (though not very likely) that the Cuckoo's superficial resemblance to a hawk may intimidate the birds whose nests it intends to parasitise.

INTIMIDATION

Charles Darwin thought that bright and startling colours in male birds were adaptations for attracting females. He thought that they had evolved by sexual selection; that is, that the choice of mate rested with the females, and that those males which were most attractive to them would be most likely to get mates and perpetuate their characters through their offspring. It is probable that a great element of sexual selection is found in bird (as in most animal) evolution; but it is equally probable that the bright colours and adornments of certain birds have as their primary biological purpose intimidation and threat rather than attraction. This applies to all birds which have bright plumage, whether it be only in the male (like pheasants and some buntings), only in the female (like phalaropes), in both sexes all the year round (like Robins and Jays), or in both sexes in the breeding season (like Black-headed Gulls). Examples of the use of colour and adornment for the purposes of intimidation are:

Brightly coloured males—On the strict Darwinian view it would appear likely that the summer plumage of male Ruffs was a simple product of sexual selection. The ruff is one of the most curious waders; in the winter it is of the same build as, and not unlike, a Redshank or a Bartram's Sandpiper, but in the summer the males develop tremendous ruffs of feathers on the head and nape, which fold over the neck, and which, in full display (Fig. 35), are erected to form a circle of bright colour surrounding the head. From March to May anybody can see

35   Ruffs in aggressive display (drawing by James Fisher)

male ruffs in full plumage and display in the Waders' Aviary at the Zoo, and prove for themselves the undoubted fact that no two ruffs are alike. I know of no bird, and certainly no wader, in which such individual variation of male aggressive plumage exists, for the function of the ruffs is intimidation rather than attraction. In the breeding season Ruffs congregate at special meeting-places, usually called Ruff "hills" or "leks". On these hills the males display and threaten with fully expanded ruffs and ear tufts, not at the females but at each other.

Battle is much too symbolised to allow much loss of blood or feathers, but occasionally real scraps will develop. The males crouch and face each other, beak to beak and with ruffs spread, or chase each other to and fro from the little areas of the hill that each seems to call his own territory and defend as such. It appears that bluff has been largely substituted for beak and claw fighting, a bluff accentuated and assisted by the brightness and extent of the ruffs. The females (Reeves), meanwhile, wait quietly in the wings. She certainly appears to select her mate (for the males are completely promiscuous), and walks over to him, crouching in a submissive posture and touching him with her beak before mating takes place. The female after mating sees no more of the male. She selects the nest site, builds, lays and raises the family. Females disperse some distance away from the leks to build, which is an additional defence against predators.

Blackcock breed in certain parts of Britain (unlike Ruffs, which were only known to have bred once between 1918 and the sixties, but now breed most years). Like those astonishing waders, they have meeting-grounds (also called leks) on which they congregate. At these meeting-grounds male Blackcock have three kinds of display—first, crowing and jumping, which appear to be a mutually-stimulating affair; second-ly, the highly aggressive "rookooing" in which, to use Lack's descrip-tion, "the male stands with head and neck thrust forward, the neck enor-mously swelled, and utters a musical, bubbling dove-like 'rookooing', often in long continuous phrases, the whole body shaking with the calls," and at the same time the lyre-shaped tail is spread, the wings are dropped to show patches of white, and the red waffles fully dis-tended—as they are in all forms of the Blackcock's display; thirdly, the circling, crouching courtship which often ends in copulation with the female.

At Blackcock leks the males all hold small territories to themselves; the object seems to be to prevent other birds from interfering with the male when he is copulating with the female. The whole system of the Blackcock's sexual behaviour centres round this territory; were it not efficiently secured and defended, he would have little sexual success. Hence the importance of his striking plumage, his use of it in display, and his song (for this is what "rookooing" is). That his plumage is used in a socially stimulative display, and that it is used in a courting display directed towards the female, once she has walked into his territory, seems to be a secondary matter. The Blackcock's intimidation display is not necessarily all bluff; sometimes mortal combats take place. This is wasteful, of course, in the biological sense, but seems to be quite rare once the male has established his territory.

The Snow Bunting is a bird which nests sparsely in the north of Scotland, and widely in the Arctic. It is a monogamous bird (though oc-casional polygamy has been noted), and the male in breeding plumage, with his attractive black and white colouration, contrasts strikingly with the brown and white female. Though his contrasting colouration is brought into use in special display ceremonies devoted only to his female, its primary function appears to be one of threat (see Fig. 36). When the male has arrived in his breeding territory in spring and notices another Snow Bunting, whether male or female, approaching his territory, he goes back at once into a threat attitude. He faces the in-truder, lowers his head between his shoulders, and utters a sound which Tinbergen writes as pEEE, and likens to the sawing note of the Coal Tit. If the newcomer alights in his territory he flies towards it, singing in flight, and rising steeply with his body curved upwards and his wings trembling. Only when he is near the other bird does he distinguish, by its reaction to his approach, between male and female. If the intruder is a male, a struggle may result; if a female (and he is unmated), a courtship display. Here then is a case of a first general reaction being entirely one of threat, continued as such if the other bird is recognised as a male, and substituted by a courtship display only when, at close quarters, the newcomer is recognised as a female.

To observe Ruffs in any numbers in natural surroundings it is necessary to go to Holland; Blackcock leks are few and far between; Snow Buntings take a lot of finding on remote Scottish hilltops; we must therefore find another example a little nearer home. Let us examine the Reed Bunting as described in Eliot Howard's *An Introduction to the Study of Bird Behaviour*. Eliot Howard (who will be mentioned a good deal) has given us a picture of its behaviour whose accuracy of description is only equalled by its quality of prose. Though we cannot quite

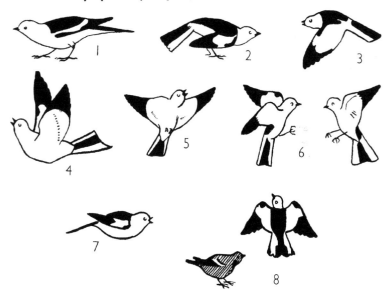

36 Snow Bunting—display attitudes: 1, 2 threat; 3, 4, 5 song flight; 6 territorial fight; 7 calling on arrival in territory; 8 display to newly arrived female (drawings by James Fisher)

compare his account step by step with Tinbergen's description of the behaviour of the male Snow Bunting, we can find in it enough to tell us that the male Reed Bunting, no less than its arctic counterpart, lives a spring life in which intimidation, aggression and territory occupy its mind to the exclusion of almost every other instinct. Intimidation (Howard calls it excitement) is expressed by voice, flight, and movement at rest. Song, the chief signal of the possession of the territory, doubles in its intensity at the approach of another bird. Sexual flight seems to be of two sorts—one like a butterfly—slow flaps with full wings—the other like a moth—rapid vibrations of partly opened wings (compare the threatening flight of the Snow Bunting). Perching on a branch or the ground, the male flutters one or both wings, or expands his tail. These displays accompany attack on a territorial rival, but when the female is at close quarters they are transferred into what a dancer might call the "routines" of a chase that may end in successful or unsuccessful mating. We cannot go very far wrong if we interpret his general reaction as being one of threat (like the Snow Bunting) and his special reaction battle or courtship, depending whether, at close quarters, he finds a female or another male.

Brightly coloured females—Some birds, like phalaropes and Button Quails, have brightly coloured females and relatively dull-coloured males. The roles of male and female in the life-cycle seem to be almost completely reversed; when the female has laid the eggs it is the male which has to sit on them. It is not surprising therefore to find that, among these peculiar birds, the female takes the initiative in marking out the territory, intimidates other females, and uses her bright colouration in male-like postures.

Bright colours in both sexes—If it is true (as we shall see later) that the sexes are often recognised by their posture and movements, even among some birds in which the sexes have different colours and adornments, how much more true it be in the many cases where male and female have almost identical plumage. In the breeding season the male Goldcrest has a somewhat more fiery golden crest than the female—otherwise they are outwardly identical. In an encounter between rival males the crests are erected and used as the special signalling organs of threat attitudes. They are used in courtship also, but in this case the erection of the crest by the male evokes a different response from the female. Much the same situation can be seen when watching the use of their contrasting colouration by Great Tits or Blue

Tits in sexual rivalry or courtship.

Robins, as David Lack mildly put it in his superb and intimate study *The Life of the Robin*, are "renowned for their pugnacity". Without any doubt they are the most highly aggressive of any British bird. Intimidation is their daily life. Bright colours, with an entirely aggressive purpose, are found equally in both sexes and, except in part of their breeding behaviour, so are intimidation and aggression. In autumn and early winter both sexes hold territories which all except a few of the females defend by song and by aggressive posturing. The male continues his territorial and aggressive activities practically throughout the year, being joined in the spring often by a female from a neighbouring territory, who then ceases to be a rival and becomes a mate, ceasing also her songs and aggressive display.

Robins posture by stretching throat and breast to display the maximum amount of red, and accompany this by swaying from side to side. Colour, posture, and song are important. Robins will attack the red breasts of whole stuffed adults (Plate 10), or just the red breast alone stuck up on a piece of wire (in one case, after many attacks, a mount was taken away and the robin continued to attack the place in the air where it had been), or even a stuffed bird with the red of the breast covered with brown ink; Robins also pursue other Robins (or even other species of birds) when they see them flying away; they also reply to song by song, or to a distant view of another Robin by song. Their mates alone, whom they recognise individually, are immune from this sort of treatment.

Bright colours in both sexes in the breeding season—Many kinds of birds have a seasonal change which is of the same nature in both sexes. In winter the Black-headed Gull has a white head (with small dark markings) and in summer a rich chocolate-brown one in both sexes (this hood is never black). It seems that this donning of summer plumage in both sexes is found very widely in social birds, like some gulls, Guillemots, and Puffins; less often in non-social kinds.

In a Black-headed Gull colony both birds of the pair take a roughly equal part in all the business of breeding, and both defend, by means of threat, a small area round their nest. Hostility will be shown by a nesting gull of either sex only when the territory is invaded—gulls will tolerate each other in the neutral ground away from the nests. By using stuffed birds we can find what it is in the intruder that causes the hostility (Fig. 37). A stuffed Black-headed Gull in full breeding plumage, if put in a

37   Black-headed Gull attacking stuffed birds in winter (left) and summer plumage (right)

territory, is vigorously attacked by the owners not only with their beaks (with which they attack live intruding gulls), but also with their feet. Normally, the Black-headed Gulls only attack human intruders with their feet—it can be imagined that the stuffed mount is the only sort of invader likely to wait until the defender can get its feet into action.

Most of these attacks are directed against the "black" head of the mount; clearly this adornment is important. We can find, by experiment, the limits of its importance. If we set up another stuffed gull, this time in winter plumage—with a white, not chocolate, head—we see that it is attacked almost as vigorously as the mount in full breeding plumage. If we set up a corpse of a gull it will be attacked as long as its pose bears some relation to that of a living bird. If we destroy the corpse's realism by cutting off its head, for instance, it will no longer be attacked, though the gulls may be apparently puzzled and frightened by it.

So we have established that the black head is not all-important in provoking intimidation, but that pose is important too. But the black head seems to focus the display; and, even by itself, it can "release" the aggressive display of the territory owner, provided it is in the right position. Set up on a skewer at normal height, the head of a decapitated bird is attacked with beak and feet. If it falls to the ground, on the other hand, it seems to be no longer capable of releasing aggressive behaviour; its position is now more than that of an egg of an intruder. In fact, sometimes when this happens the living bird may roll the head into its nest and incubate it!

SUBMISSION

The complicated system of postures, movements, and the use of colouration and song represented by aggressive display has a definite biological purpose. It is of use. In their sexual rivalry, birds seldom fight to the death; rather do they play a game of chess in which their territory is often the chess-board, and their postures, colours and adornments the pieces. The competitor or intruder nearly always "knows when he is beaten" before feathers or even tempers are lost. Clearly this system is a highly economical one—the loser is unhurt and free to carve out a territory elsewhere, or to try his powers of bluff on a different male. The species as a whole is prevented from suffering from the waste of its precious individuals. As we will discuss in the next chapter, the territorial system is of such usefulness and importance that birds as a whole can afford this complicated (if not individually fatal) system of sorting themselves out in the breeding season.

In certain kinds of birds it seems to be necessary to have additional devices to avoid actual battle or after a display of intimidation. This is particularly true of social birds. Among birds which normally live in flocks, and particularly among those in which flocking extends into or even through the breeding season, the normal releasing of sexual rivalry might lead to serious complications were not some such system in existence. Thus we find that Jackdaws, when threatened by others of their kind, may turn and expose the least protected part of their bodies—the backs of their necks. In the Jackdaw this is even specially marked by a grey patch. When this submissive action is performed, the aggressor is assured of moral victory, which in a sense is all he wants. Submissive attitudes have been recorded in several other social birds (the Night Heron, for example), and, now that their importance is realised, they are being found to exist quite widely in many sorts of animals.

The importance of submission is shown in an interesting way by the behaviour, in close confinement, of such male birds as do not submit. (In nature these move off somewhere else after an exchange, which they cannot do in captivity.) When several birds are confined together, one soon becomes the despot, and when he is removed, another dictator takes his place. In fact, such birds can be arranged in descending order, according to which pecks which. Four male Blackbirds, which Tinbergen kept in a small cage, "developed a severe despotism, as captive birds are apt to do in the spring, one of the males actually killing all the others." This certainly would not have happened had Blackbirds been naturally social birds, for then it is probable that some submissive device would have been developed to prevent such mortal combats. It is interesting to note that the domestic fowl, which in nature is not a social bird, has, in the course of some thousands of generations of captivity, developed a system by which it can hold a painless social hierarchy, and that submissiveness plays a large part in preserving peace in the farmyard. That this does not always apply can, however be deduced from the continued existence of fighting cocks.

8 How birds recognise one another

"I have now, past dispute, made out three distinct species of the willow-wrens (Motacilla trochili) which constantly and invariably use distinct notes. But at the same time I am obliged to confess that I know nothing of your willow-lark. In my letter of 18 April, I had told you peremptorily that I knew your willow-lark, but had not seen it then; but when I came to procure it, it proved in all respects a very Motacilla trochilus, only that it is a size larger than the other two, and the yellow-green of the whole upper part of the body is more vivid, and the belly of a clearer white. I have specimens of the three sorts now lying before me, and can discern that there are three graduations of sizes and that the least has black legs, and the other two flesh-coloured ones. The yellowist bird is considerably the largest, and has its quill feathers and secondary feathers tipped with white, which the others have not. This last haunts only the tops of trees in high beechen woods, and makes a sibilous grasshopper-like noise, now and then, at short intervals, shivering a little with its wings when it sings, and is, I make no doubt now, the Regulus non cristatus of Ray, which he says, 'cantat voce stridula locustae.' Yet this great ornithologist never suspected that there were three species."

GILBERT WHITE, *17 August 1768*

We must avoid placing ourselves in the birds' position when we deal with the methods by which they recognise one another, whether others of their own or the opposite sex, their parents or offspring, or their species. We have no philosophical justification for imagining that, because we recognise differences between a male and a female bird or between two different sorts, the birds themselves recognise these differences by the same characteristics. Only by a long process of objective observation, by taking nothing for granted, and often with the help of actual experiments in the field, can we find out the real basis of recognition in birds. It is becoming clear, as has already been hinted, that in many cases postures and movements are just as important as, or even more important than, special colouration in sex recognition.

The male Flicker (a kind of American woodpecker) is distinguishable to man from the female, since the male possesses a handsome "moustache" of black feathers on each side of his chin. Apparently a male Flicker can also distinguish a male from a female by this characteristic, as has been proved by some recent field experiments. A pair of birds had settled down to normal breeding behaviour, and were clearly recognising each other by sight or habit or by posture and movement, or by a combination of these. The female was then provided with a false moustache. When she was released the male approached her from behind and began to mount her (the pair had reached this stage in their breeding cycle). She turned round and he saw her moustache. Immediately he went into his full "anti-male" aggressive display, pursuing the unfortunate female for two and a half hours and spreading his tail to show the bright under-surface.

This experiment shows, first, that Flickers can recognise the sex of other Flickers by posture, secondly, that the moustache, an adornment perhaps, to human eyes, is also a recognition mark, thirdly, that the contradiction artificially produced between male (moustache) and female (posture) characters produces great excitability and a generally anti-male reaction; and finally that the exhibition of the bright under-surface of the tail during the chase may be (in its colour and the way it is shown) a further male-recognitional character.

In the Snow Bunting, Dr Tinbergen found that it was mainly the attitude and movements of the bird against which threat display was directed that determined whether the threat should be continued as such or transformed into courtship display. This explains some of the occasional cases of apparent homosexualism in birds. If, by some upset of its sex-glands, or another cause, a male bird goes into a female attitude, it will often be accepted as a female even though it may bear male plumage—the displaying male may even try to mate with it. Among the Ruffs and Reeves in the Waders' Aviary at the Zoo there are many more males than females, and some of these often take up female attitudes. The consequent attempts at mating have often led onlookers to suppose that Reeves also bear Ruffs.

Observers of the sexual behaviour of birds in the field have often (as described in the case of the Black-headed Gull) used stuffed birds to test reactions and recognition. In some cases this has been the means of discovering how the nature of a bird's sex is recognised by a rival or by a possible mate. If a male Ruffed Grouse (an American bird) is presented with the skin of a male of its own species stuffed in an ordinary attitude, it will try to copulate with it—showing that in this particular case it is the attitude, rather than the characters or adornments, that is important in recognition—since the attitude in which a bird is normally stuffed is essentially female, having, as Dr Tinbergen points out, the lack of movement and rather hunched position of a female willing to copulate.

The Blackcock does not show the same reactions to stuffed mounts as does the Ruffed Grouse. This bird (in which the sexes are more widely different) always recognises the sex of the dummies; males attempt to copulate with stuffed females, strike at stuffed males in the normal position, and, if a male is stuffed in the "rookooing" attitude, may strike repeatedly at the adornments of the head. One bird eventually removed the head and continued to strike at it where it lay three feet from the body.

We have now discussed two mechanisms, important in the recognition of the sexes—colour (and adornment) and posture. Male birds may distinguish males from females by one or both of these characteristics. There remains another important mechanism used for the purpose of recognition—voice and song. Vocal sounds play their part in enabling birds to recognise one another individually. Recognition, by birds, of individuals of their own species is a fact, though not a universal one. A lot of work is being done on this subject; among recent discoveries are: Black-headed Gulls and Song Sparrows recognise their neighbours, but not strangers; paired Willets (an American wader) can recognise each other without a display; gulls and penguins recognise and greet their mates from afar; immature Night Herons (paired) recognise each other after 20 days' separation, but not after 6 days if extra feathers are glued to the head of one; young Night Herons do not recognise their parents, but young Herring Gulls do, and their brothers and sisters as well; most social birds recognise their own young. Gentoo Penguins recognise their mates by crowing, though they sometimes make mistakes.

Song is a very important factor in enabling certain species which are very similar in plumage, and which overlap in their breeding range, to distinguish each other. In Britain we have two examples of pairs of species which are almost indistinguishable in size, colour, and plumage: the Chiffchaff and Willow Warbler, and the Marsh and Willow Tits. Each pair of birds is very closely related, from the evidence of their anatomical structure, and it can be imagined that it was not very long since, in evolutionary perspective, that each pair shared a common ancestor.

Without going into the arguments about how these species of birds have come to have the same geographical distribution, we can examine the devices by which they are able to distinguish each other, and so keep their identity as entirely different species. In both cases, warblers and tits, the important mechanism seems to be song, or at least voice. The difference between the "song" of the Chiffchaff (harsh and disyllabic) and the trill of the Willow Warbler (sweet, plaintive note—Gilbert White) is clear, so much so that it it is the chief character which enables the human being, investigating the bird life of woodland, garden or scrubland, to tell them apart. Though their habitats are not quite the same—Chiffchaffs like older woodland with high song-posts, while Willow Warblers seem to prefer open scrubland and can tolerate low song-posts—there can be no doubt that the difference in song aids the birds in avoiding a serious mistake in identification. The Wood Warbler, it might also be said, is outwardly similar to those two, and also has a very distinctive song (sibilous, shivering noise—Gilbert White).

Like these warblers, the Marsh and Willow Tits overlap in broad geographical distribution, and a good deal in habitat. Although they are superficially so much alike that ornithologists did not separate the British species until 1900 (the chief anatomical difference between them lies in the minute structure of the feathers of the head), yet they are able to distinguish their own species, and thus avoid wasteful attempts to interbreed, by their call notes and songs. These are absolutely distinct. The call notes of the Marsh Tit are four, which can be written, *pitchuu* (the commonest), *tsee-tsee-tsee*, *chick-a-bee-bee-bee*, and a harsh note *chay*. The last is the only one which is at all like the *chay-chay-chay* of the Willow Tit, which in any case is much more harsh and twanging. The Willow Tit never uses *pitchuu*, though it may make a sound *chich* in its place, and its thin *eez-eez-eez* or high *zi-zit* are not found in the Marsh Tit at all. The songs again are entirely different in the two species—the *piu-piu-piu* of the Willow Tit is quite unlike the *schuppi, schip,* or *pitchaweco* syllables of the Marsh Tit.

Practically all the finest songs of British birds are those in which the sexes are alike (the Blackbird and the Blackcap are exceptions). Though the main function of song in most perching birds must be regarded as signalling the possession of a territory, we are prompted to believe that the quality, extent, and varied nature of these songs also form a mechanism by which the sexes recognise each other. With birds such as Nightingales and warblers, it is reasonable to suppose that a very distinctive song, in the absence of distinctive and recognisable colour, is a great help to a female in finding a mate, and distinguishing

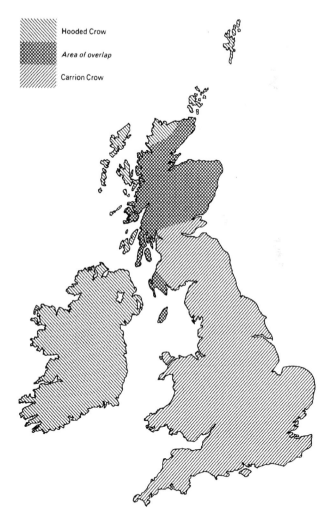

38   Breeding ranges of Hooded Crow and Carrion Crow, showing area of overlap

him when he is found.

Occasionally, in the course of evolution, a situation is produced in which such distinguishing characters as have been developed are not quite enough to keep the species entirely apart. Such a position exists with regard to the Carrion and Hooded Crows. Derived originally from a common ancestor, these birds have probably achieved, in the past, complete geographical separation, and during the period in which they lived apart they diverged from one another, one becoming (or staying) black and the other developing its typical pied plumage. Later in time, the Hooded Crow extended its bounds until once more it has come into contact with its cousin, the Carrion Crow. In Britain the line of breeding overlaps (Fig. 38), and extends from Ireland across the lowlands and part of the highlands of Scotland. Here the two birds breed in the same area, and, although for the most part they keep themselves to themselves, a remarkable number of cases of interbreeding (with fertile offspring, as far as can be seen) have been recorded. In the case of the Crows, then, it is clear that an apparent, distinguishing characteristic is not completely effective in preventing crossing. Recent experimental work with Herring and Lesser Black-backed Gulls, where eggs have been transferred from one species to the other, indicates that the chick behaves as if it were the same species as its foster parents, and when mature, will pair with that species despite obvious (to human eyes) plumage differences between silver and slate-grey mantles.

CF

Puffin

9 Territory, courtship and the breeding cycle

"During the amorous season, such a jealousy prevails between the male
birds that they can hardly bear to be together in the same hedge or field."

GILBERT WHITE, *8 February 1772*

A THEORETICAL CASE

There are two modern definitions of bird territory which it will be well
to keep in mind. A general one (Tinbergen) is "whenever sexual fighting
is confined to a restricted area, that area is a territory." A more special
one (Lack) describes a territory as an "isolated area defended by one individual of a species or by a breeding pair against intruders of the same
species and in which the owner of the territory makes itself
conspicuous."

We must also bear in mind that the machinery of keeping a territory
is based on the aggressive colouration, adornment, song and action
which we have discussed in the last chapter. With this definition and
condition we can attempt to describe a bird's territorial life. The bird we
will select is imaginary, though there are many living birds like it; imaginary because no one species of bird shows all the typical attributes of
a territory-holder, and because it will be to our advantage to discuss the
rules before the exceptions, the type before the examples.

Our standard bird, then, is a land bird—a passerine (or perching
bird); it inhabits hedgerows, gardens, scrubland, or woodland; has a
brightly coloured male and a dull female; has a male which sings; is
partly migratory and partly resident; has catholic tastes in food; is, on
the whole, monogamous. In winter mixed flocks of males and females
work the hedgerows, fields, gardens, wood-edges, rickyards, or beaches
above high-water mark of some country or countries in the Holarctic
Region—temperate Europe, Asia, or North America.

At the end of the winter—perhaps in early March—these flocks
cease to be the entirely amicable affairs they have been since autumn.
They still retain their function—that of co-operative food finding, so
that the discovery of one is shared by the others—but quarrels, not over

food so far as can be seen, nor over anything tangible, begin to break out between males. They may flirt their wings or posture at one another, and have little fruitless chases—nothing serious, but enough to show that some internal change is taking place in the males' make-up that prevents them, temporarily, from behaving like ordinary members of the flock.

As time goes on these quarrels tend to become slightly more serious and more prolonged, and are accompanied by little expositions of the breeding song, but before they become really acute the males begin to go away for short periods instead. They pay visits to suitable small areas of their summer breeding habitat. At first these visits are very short, and involve the prospecting of a fairly wide area of the summer habitat; as time goes on the period spent away from the flock grows longer and longer, and the area visited smaller and smaller, until a day comes (or a night rather) when the male has left the flock altogether and sleeps in what is now the "rough copy" of his breeding territory.

By now he is well acquainted with his home. The boundaries of it interest him little as yet; most of his business concerns several points of vantage, to which he continually hops or flies, and from which he sings first a few quiet bars, or a little shadowy copy of what, in a day or two, he will be pouring forth with all his might as his full spring song. The paths to his songposts become as clear to him as the route from garden gate to back door is to us. The post or gate or branch from which he sings is approached perhaps by a flight to a bush, another to a branch, a hop to the left, a quick flight to the right, and a hop straight upwards: he seldom departs from this routine.

As time goes on he visits his headquarters more often, and may sing for as much as an hour on end; he seems to favour one or two strategic songposts more and more, so he cuts the number down, though, when passing the immediate neighbourhood of one that he has discarded, he may still go through his stereotyped routine of path-movement. With the increase of his song and preoccupation with headquarters he goes less far afield for foraging, until he is feeding entirely in the loosely defined area he owns.

This area does not stay loosely defined for long. Other males of the same species, who have also left the flocks, approach, fly past, sing, and come foraging in his area. He is not tolerant of these birds. He chases them, postures aggressively, and sings at them. Soon he finds that certain of these birds are his neighbours, staking claims on the other side of the garden, on a branch of the next big tree, or fifty yards down the hedge. His aggressive reactions towards them are more patterned; he plays them at song-tennis over the neutral ground between—ground which rapidly becomes less and less neutral. When one of his neighbours, that are now his rivals, alights in his area, he makes a display flight at it, often singing on the wing; the neighbour retreats into his own area, and if he is followed, the rôles are reversed. In such a way the boundaries of the territories are marked out. Eventually it becomes possible for the human observer to plot the borders of the territories on a map, almost down to the last blade of grass. Over these borders are fought the most bitter battles—here there is least to distinguish the rivals in pugnacity. For our birds recognise the aggressor, and the further a fight ranges into the home bird's territory, the more aggressive does he become, and the less aggressive his rival; only on the borders are they equally matched—in fact, the border *is* where they are equally matched.

When our male first became resident in his territory, he owned about two acres. His battles with his neighbours have cut the area down to one; he may now be in relation to four or more other males on different sides, each with about an acre, too. It will be perhaps about three weeks since he left the flock.

It is now the females' turn to leave the flock. They seem to forage individually farther and farther afield; one passes through a male's territory and is threatened as if she was a male; her time is not quite ripe, for she takes no notice of his song and goes on her way. The next one that arrives has been attracted by the male's song—she lands in the territory and approaches. The male greets her in exactly the same way as he would a rival male entering his territory—by threat. Not until she is at close quarters does he recognise her and change his attitude. Then comes a sexual chase. Round and round the territory the two birds fly, the male trying to grasp the female's back, bring her to the ground, and mate with her. Eventually he gives up, and both land, the male often finishing with a little courtship display, and often a call that seems to represent frustration. For, though the male is ready to copulate, the female is not, and may not be for some days or even weeks—although she was ready to be attracted into the territory in the first place.

Sexual chases now become established as a sort of daily routine, and though they indicate that successful copulation has not yet taken place, they show all the same (as Dr Tinbergen points out) that the birds are paired. Though the female will not allow the male to copulate with her, she does not leave him. Apart from their frequent sexual flights, the birds learn to recognise each other individually, thus dispensing more and more with the usual recognitional displays. The female helps the male to protect the territory (and his aggressive attitude to his neighbours increases still more), not by attacking his rival males, but by attacking other females.

At this stage the male's aggressive behaviour is still increasing, and is strongest towards the centre or headquarters of the territory, and less strong towards the borders. Even now, when neighbouring males are mated, and consequently more pugnacious, a new male, late from the flock, may (no doubt encouraged by the absence of song) try to stake out a territory between those already occupied. Sometimes he may succeed, depending on the distance between his chosen headquarters and that of the birds already in possession, for up to a certain minimum limit territories are compressible. If he does not transgress this limit, he may succeed; if he does, he gives up the struggle and moves on.

One day the female shows great interest in one or two bushes, or a stretch of hedge. She picks up a scrap of nest material. The male approaches her, and this time the normal sexual chase does not follow. She is ripe for copulation, the state of her breeding organs has crossed a certain threshold; with flattened back and lifted tail she invites the male, and he mounts and copulates with her. But even now this does not always happen—she may invite, but the male may not be receptive, and then she will relinquish her attitude; the male may invite, and another sexual chase take place; copulation may be unsuccessful, and may result in courtship posturing. Rapidly, however, the birds become used to one another, and the female starts building the nest in earnest. The male too shows signs of building activity; he may pass building material to the female, may add it to the nest himself, or may make feeble and unfinished "cock-nests" of his own. From a week to a fortnight after the first copulation the nest is complete.

The female now becomes restless, and makes frequent visits to the nest. One morning, early, she visits it for over half an hour; she does this again the next day and the next and the next. Four eggs are there when we look, though during those four days of laying the nest was still being lined with feathers. The female no longer allows the male to copulate; for his part he resumes his song, which is almost as strong as if he was unmated, though he has been relatively quiet since she arrived in his territory. A day or two after she has laid the fourth egg she begins to sit. The male takes no part in the incubation, but continues his territorial song and aggressiveness in the intervals of foraging for himself and taking food to the female on the nest.

After a fortnight the eggs hatch. The young are like those of most passerine birds, at first almost naked and blind. The business of feeding them occupies the parents almost the whole of their time. For a day or two the young are fed from the crop; after this large quantities of insects are needed and are brought to the nest three or four times an hour. As the nestlings grow larger and hungrier, every hour of daylight means five or six, then eight or nine, then twelve, and, towards the end, sixteen visits. To collect these insects the parents have to forage wherever insects are, and parties of parents will hunt together in neutral ground or even over part of the territory of one of the pairs, and even in the latter case little aggressive behaviour will be aroused. Every day the young grow noisier and more active. A fortnight after hatching their excrement is no longer voided in a little gelatinous bag which can be removed by their parents, but is deposited instead on the rim of the nest (a sign, by the way, of a successful nest if you come across it later in the season). There is no longer need to avoid fouling the nest; though they cannot fly properly they have left it. They hide now, a special call signalling their presence to their parents. Sometimes the parents each take charge of half the fledglings. After a week or ten days the young begin to feed themselves, after another week they are no longer fed by their parents at all, in another day or two their parent-calls have been replaced by the ordinary adult flocking-call, and they are off on their own.

Long before they finally go altogether the young have been wandering more or less where they would. Even when fed by their parents they have trespassed into neighbouring territories where they, and often their parents too, have been tolerated. But when they are finally independent their parents are found to be defending their own territory again; the male, who was too busy to sing while he was feeding the young is in full song, the female refusing copulation until she is ripe for the recurrence of her cycle that means a second brood. And after the whole process has been gone through again with the second brood, the parents (unless indeed they have a third or fourth) will lose their urge to copulate (an urge which, we have seen, is of longer and steadier duration in the male than in the female). Moult will have started—the old (and by now very worn) feathers falling out in strict sequence over perhaps two months, and being replaced by fresh new ones with good insulating power to stave off winter's chill night winds. The birds will lose their territorial sense; flock communication sounds will replace male song; and, like their young, they will wander off foraging and the autumn flock will be established again.

The composite bird which we have described in such detail owes much to the buntings and to those who have so carefully studied them. Now we have a picture of the life of a territorial bird, and we can compare the lives of real birds with this picture. We have traced the breeding history of our standard bird stage by stage from the moment when the winter flocks break up until the moment when the autumn birds show signs of flocking again. On our way we have passed several milestones; let us take the journey again now, resting for a time at each milestone, and see what the different sorts of birds are really doing.

TAKING UP TERRITORY, AND WINTER TERRITORY

Territories can be taken up by males (the most usual case, as with our example), by both sexes together (particularly among seabirds and some arctic waders that pair up on migration, before they have arrived at their breeding place), or by females (for instance, phalaropes and Button Quails, where the female has the bright colours and the aggressive display). To take up the territories, if they are a migratory species, the males (we shall exclude the other cases) leave first for the north ahead of the females; or stay more or less where they are, if they are resident or hold winter territories. Birds which do the latter deserve rather a special mention. If we look back to page 133 we see that Tinbergen's definition of a territory hints at a sexual purpose, while that of Lack (who extensively studied the Robin, which holds a winter territory) carefully avoids doing this. If we look—from an evolutionary point of view—at the business of winter territory, in which males and females which were mates in the summer may be rivals in the autumn, we can see that these two definitions may not be contradictory, provided, as seems likely, one habit has evolved out of the other. It is possible to imagine that the keeping of winter territory by such species as practise it may, in the course of evolution, have been an extension or resumption of the normal breeding-cycle territory, and that though originally a purely sexual affair, it now serves in winter for other purposes. For instance, it keeps the species evenly spread in regions of scanty food, but of course the effectiveness of this depends on the kind of food requirement. Many invertebrate-eating birds derive advantage in winter flocking, since a group or swarm of hibernating creatures turned up by one bird becomes the prey of all.

Winter territory persists, to a small extent, in quite a number of birds. Those in whose lives it plays an important part range from California shrikes and Mocking Birds to San Francisco Spotted Towhees; in Britain the male Blackbird holds a winter territory to a certain extent; and some females hold them also. The finest example in Britain, though, is the Robin.

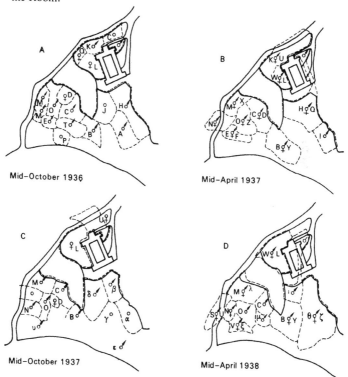

39 A two-year history of Robin territories (after David Lack)—letters, roman or greek, represent individual birds; circles indicate occupied territories whose boundaries are the dotted lines. An individual's territory and the history of its mating can be traced on the sequence of maps

A Robin may inhabit only a few acres during the whole of its life, though certain females wander, and a few are even migratory. In winter (contrary to the impression given by some Christmas cards) both male *and* female hold individual territories, which *together* are almost exactly the size of their summer ones. These they defend, like most aggressive birds, by song and dance. In late winter or spring the females cease their territorial activity, and may often mate with a neighbouring male rival, sometimes the same mate in successive years (Fig. 39). Only in early autumn, after the waning of breeding activity, is there a time-gap in which territorial activity is low.

There is only one bird known to be more highly resident than the Robin—the Wrentit, an American species belonging to a family of birds not occurring in the British Isles. A male Wrentit takes up a territory in the first March after he has fledged, and keeps it all the year round for the rest of his life. He stakes it out and keeps it by song and fight, and when a female joins him she helps in the fighting. If the female dies, the male waits for a new one; if the male dies, the female deserts the territory for, or fuses it with, that of an unmated male. Such territories as fall vacant are taken by males (mainly young) that have none.

It appears that both our definitions of territory (p. 133) work in the case of the Wrentit. To judge from their behaviour, sexual activity in the pair never falls below a stage equivalent to the period of sexual chases in our standard bird—that is, the male and female *are bound by a sexual bond throughout the year*, though they only reach the stage where the female will accept copulation at breeding time. This behaviour might seem ordinary to a student of monkeys or man, but is quite remarkable among birds.

STAKING THE CLAIM

Practically all territorial birds use song, chasing, and fighting in the defence of their plot of ground. In the buntings (which include the American "sparrows"), behaviour (see Fig. 36) in defence of the area (see Fig. 40) is almost identical with that of our standard bird. The key to claim-staking is *exhibitionism*; birds make it obvious to others that they are in possession. Hence species with weak songs, Stonechats, for example, have a headquarters where they stand in full view and show their contrasting plumage; at the opposite extreme, Nightingales, which make no appeal to sight, have a correspondingly loud song.

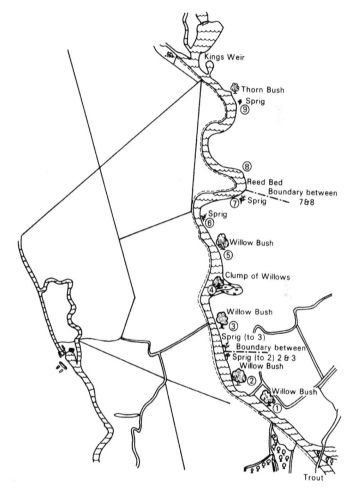

40   Reed Bunting territories on the River Isis between The Trout and King's
Weir. Nine male bird territories have been plotted from observation of song
posts. Note the even spread of territories, conditioned, even so, by the supply of
sprigs and bushes to serve as song post along the three-quarter of a mile stretch
of river

Actual fighting is replaced by bluff and other "substitute ceremonies" in many species, particularly social ones, where fighting would be an expensive and useless luxury. Thus Herring Gulls do not fight, but symbolically pluck grass, and Black-headed Gulls seldom fight with an immediate neighbour in their colony, but only seriously with strange birds not known to them.

Provided they can recognise them as such, territorial birds do not as a rule attack birds of species other than their own. Although Snow Buntings have the same threat reaction to males and females of their own species when some way away, and thus *may* not distinguish between them at a distance (although their plumage seems very different to us), yet they can distinguish their own kind from Lapland Buntings or Wheatears, which often share the same breeding-ground. They have never been seen to threaten these other species. Though these other birds are to a certain extent competing for the same food, they are not sexual competitors, and to a breeding Snow Bunting it is sex that matters.

In the same way breeding Willow Warblers and Nightjars threaten their own species (i.e. sexual competitors), but not others which eat the same food. Nuthatches, however, display to and attack birds of such other species as are competing for nest-sites, like tits and Starlings; but these battles are against birds which might interfere with their breeding-cycle, not their food supply.

Swans will chase any *large* animal form their territory; although Ringed Plovers will chase Skylarks and Linnets, Little Ringed Plovers (a smaller species) will even try to chase sheep. We can perhaps explain this by pointing out that these birds are guarding the nest; Ringed Plovers, with their nest on the ground, have it exposed to the danger of being trodden on by mistake as well as deliberately raided—not that this explains their attitude to Larks and Linnets.

We can thus reach the general conclusion that birds normally attack other species in their territory only when those animals tend to interfere with the breeding-cycle.

SIZE AND SHAPE OF TERRITORY

The normal, average size of bird territories is greatly variable. For non-colonial and easily studied species (David Lack's Robins or David Snow's Blackbirds) this often appears to be between a half and one-and-

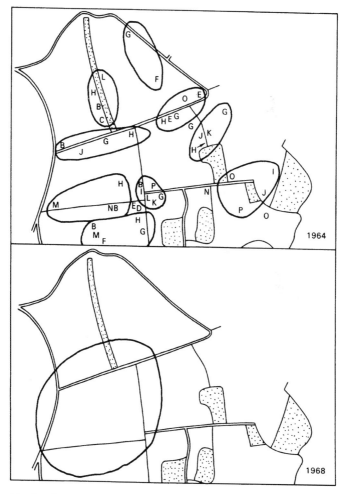

41   Partridge territories on Suffolk farmland in 1964 and 1965 from CBC reports, showing decline in territories due to changes in agricultural practice and use of pesticides, with consequent reduction in insect food for the birds

a-half acres. However, Kenneth Williamson, looking at Wrens on farmland, found that whilst the size (and shape) of territories depended on the nature of the habitat (i.e. territories were more or less circular in woods, but very elongated in hedgerows), the temporary reduction in Wren numbers and subsequent rapid increase following the severe 1963 winter made little difference to the size of territory occupied. What happened was that when all the available "first-choice" habitat (woodland for the Wren) had been filled, then territories began to be taken up in "secondary" habitat like hedgerows, and in this poorer ground they were larger and held in less stable tenure (Figs. 41–43).

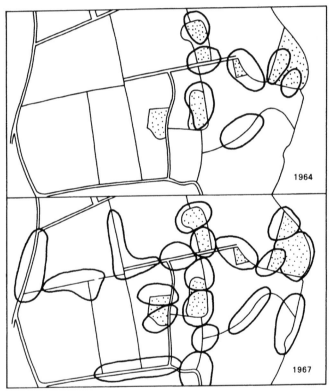

42   Wren territories in Suffolk, 1964 and 1967, from CBC reports, showing recovery of the species after its heavy losses in the severe winter of 1962/63

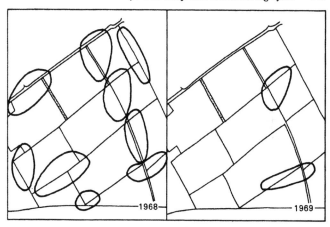

43 Whitethroat territories in Suffolk 1968 and 1969, from CBC reports, showing results of population "crash" between the seasons

Clearly, for colonial birds (the gulls, or Gannets, for example) only the immediate vicinity of the nest is protected, and this is true for some of the ducks, which may also have a second, separate, "territory" on their main feeding water.

SONGPOSTS, NESTS, AND FEEDING-GROUND

Our standard bird had a few very definite songposts in his territory, from which he sang and displayed. As time went on he reduced the number of these, showing a clear preference for one or two. This is not always so. Some of the American warblers never have definite songposts, but sing from anywhere within the territory. The Rosy Finch seems to use not a bush, a branch, or a bank, but his own female. The Broad-tailed Humming-bird may have as many as eleven flight-posts in his small territory. Naturally, birds like the Skylark (which may sometimes sing from the ground or from a fence post) spend most of their song time in mid-air, over their territory but at various altitudes and positions.

The nest lay within the territory of our standard bird. Even with buntings, which approach the standard very closely, it may not do so. One of the female Snow Buntings watched by Dr Tinbergen built her nest some distance outside the male's territory. By dint of a two-day "battle" with the neighbour, he conquered the ground near the nest. Other buntings (like Bachman's Sparrow in America), and even warblers such as Chiffchaffs, may sometimes build nests outside the territory, as much as 100 to 150 yards from the territorial headquarters. African Bishop-birds, which are polygamous, have three or four wives, and the nests of one or more may occasionally be outside the territory.

Much discussion and controversy has taken place as to the biological value of territory. In particular, the relation between territory and food supply has occupied many pages of print in the learned journals. Howard was the first to suggest that the system of territory holding might confer advantages on birds by assuring a regular food supply. Some critics have maintained that because birds did not always seek food in their territories, this was not so. It will be remembered that our standard bird sought food for its young outside its territory; as far as I know there is no single case yet published of any species of bird which is *never* known to seek food outside its territory, during such time as it holds one. Warblers, buntings, and Dabchicks seek food *mostly* in their territories. American Robins and Purple Sandpipers feed in them, but also at a common feeding-ground where they do not attack neighbours. Common feeding-grounds are used by Lapwings, Great Crested Grebes, and many other social birds, and in very few farmland or woodland areas (where most birds *are* strongly territorial) does *each* territory have a regular and reliable pool for drinking and bathing. These facilities, although sometimes lying within a territory, seem to be approached and used without an aggressive response—it would be extremely interesting to know how this is achieved.

It is clear, then, that the possession of a half-acre territory by a non-social bird confers a general advantage rather than a special one concerning the food available in the territory itself. If the latter were the case, we might expect that food-competitors of other species would not be tolerated; the evidence goes to show that they are. The general advantage lies in the fact that territory ensures an even distribution of birds over an area, thus not stretching unduly the limits of food supply in any one place while neglecting sources of food in another.

COURTSHIP

True courtship, as distinct from aggressive display, is found among all sorts of birds, territorial and otherwise. The reader is particularly referred to the pictures of courtship and aggressive display in Fig. 36. The object of courtship seems to be to strengthen the bonds between male and female, ensuring that they go through the successive phases of their cycle in a way, and at a time, suitable to one another, and, more especially, to act as a preliminary to copulation. Courtship may take the forms of vocal demonstration (grebes and divers), sexual flights and chases (buntings, see Fig. 36), and various ceremonies derived and extended from the ordinary details of daily life, such as the feeding of the female by the male (Robins, gulls, and parrots), "false drinking" (Gannets and Smews), false preening, wing-lifting (Common and Purple Sandpipers), short leap-frog flights (Green Sandpipers), or the presentation or exchange of nest material. Whilst the water dances of the grebes, and the courtship feeding of Robins and Blue Tits, are evidently essential to the pairing process and satisfying to human observers, we must not let our emotional viewpoint blind us to the possibly rather prosaic explanations. It seems very likely that courtship-feeding, for example, is valuable if the female is to achieve the right condition for egg production, and to begin the breeding season as promptly as possible. A glance at Fig. 44 shows just how much weight the female Blue Tit must put on (in terms of gonad development and the stored food to provide the nutriment within the egg) and how quickly. This becomes much easier if both birds participate in finding her food!

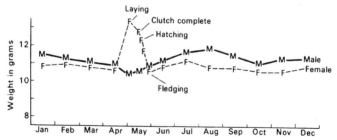

44 Weights of male and female Blue Tits throughout the year, showing spectacular weight increase in female before egg laying (from the book *Birdwatchers' Year*)

After the period of sexual flights and preliminary courtship is over, the stage of successful copulation is reached. The act may take place on

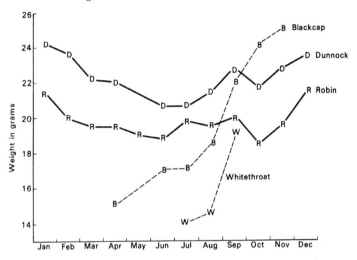

45   Weights of two resident species (Robin and Dunnock) and two migrant species (Blackcap and Whitethroat) plotted to show weight increases (fat reserves) of resident birds before the onset of winter, and before migration

land, on water (swans, ducks), or in the air (Swifts), and may follow no apparent ceremony (divers and some terns), or a special courtship display (most birds). In some birds the male may invite the female (gulls, woodpeckers, penguins), in others the female may invite the male (many ducks), either may invite either (Common Terns), or it may be difficult to decide which takes the initiative (buntings). Copulation may continue until the first egg is laid (some buntings), until incubation begins (other buntings, Herring Gulls, Skimmers), or during incubation and even after the eggs hatch (Black-headed Gulls, Common Terns).

Detailed descriptions of the courtships of all sorts of birds are now common in bird literature, so much so that the editors of the standard work on British birds (Witherby's *The Handbook of British Birds*) have managed to include a description, in many cases very detailed, of the courtship of every important bird on the British list. This book does not attempt to be a handbook so much as a guide, so there is no reason why the reader should be plagued with what has been more adequately described elsewhere. But though we must leave the appreciation of the individual details of the male-female courtship (as opposed to male-

male intimidation) of our British birds to students of the *Hand-book*—and the more students it has the better—we cannot leave the subject altogether without giving some examples of how courtship has inspired moving description in bird literature. Here are three such examples:

David Lack, scientist, on the Robin: "The first copulation and the first 'courtship-feeding' occur within a day of each other, but the two are quite separate from each other. The female invites the male to feed her with the same attitude with vigorously quivered wings and loud call as that in which the fledgling begs for food; the male comes with food in his beak and passes it to her, the female swallows it. When the performance starts in late March, the male may just hop up to the female and put food in her mouth whilst she barely quivers the wings or calls, but, after a day or two, both birds are much more excited, the female begging and being fed persistently."

Julian Huxley, scientist, with artistic approach, on the Redshank: "I spent some time watching them, and soon saw the redshanks courting. It was one of the most entrancing of spectacles. Redshanks, cock as well as hen, are sober-coloured enough as you see their trim brown bodies slipping through the herbage. But during the courtship all is changed. The cock-bird advances towards the hen with his graceful pointed wings raised above his back, showing their pure-white under-surface. He lifts his scarlet legs alternately in a deliberate way—a sort of graceful goose-step—and utters all the while a clear, far-carrying trill, full of wildness, charged with desire, piercing, and exciting. Sometimes, as he nears the hen, he begins to fan his wings a little, just lifting himself off the ground, so that he is walking on air. The hen will often suffer his approach till he is quite close, then shy away like a startled horse, and begin running, upon which he folds his wings and runs after. She generally runs in circles, as if the pursuit were not wholly disagreeable to her, and so they turn and loop over the gleaming mud. Then she pauses again, and the tremulous approach is again enacted."

Edmund Selous, artist, with scientific approach, on the Stock Dove: "However it may be, the bow itself—which I will now notice more fully—is certainly of a nuptial character, and is seen in its greatest perfection only when the male Stock Dove courts the female. This he does by either flying or walking up to her and bowing solemnly till his breast touches the ground, his tail going up at the same time to an even more than corresponding height, though with an action less solemn. The tail in its ascent is beautifully fanned, but it is not spread out like a fan, but arched, which adds to the beauty of its appearance. As it is brought down it closes again, but, should the bow be followed up, it is instantly again fanned out and sweeps the ground, as its owner, now risen from his prostrate attitude, with head erect and throat swelled, makes a little rush towards the object of his desires. The preliminary bow, however, is more usually followed by another, or by two or three others, each one being a distinct and separate affair, the bird remaining with his head sunk and tail raised and fanned for some seconds before rising to repeat. Thus it is not like two or three little bobs—which is the manner of wooing pursued by the Turtle Dove—but there is one set bow, to which but one elevation and depression of the tail belongs, and the offerer of it must not only regain his normal upright attitude, but remain in it for a perceptible period before making another. This bow, therefore, is of the most impressive and even solemn nature, and expresses, as much as anything in dumb show can express, 'Madam, I am your most devoted.' "

Nutcracker

10 What you can do

"I partly engaged that I would sometime do myself the honour to write to you on the subject of natural history; and I am the more ready to fulfil my promise, because I see you are a gentleman of great candour, and one that will make allowances; especially where the writer professes to be an *out-door naturalist*, one that takes his observations from the subject itself, and not from the writings of others."

GILBERT WHITE, *30 June 1769*

This book is about birds, birdwatching and birdwatchers. It is designed to give the reader some insight into the lives of birds and to indicate some of the fascinating realms that will open if he pursues his hobby further. Just as in learning how to distinguish one bird species from another, or how to separate the two sexes, experience is the greatest teacher, and one of birds' most welcome attractions is that they are always close to you. Wherever you are, there will almost certainly be birds and consequently, for the birdwatcher, an inexhaustible source of interest, wonderment and satisfaction. In Africa, for example, it will be the number of species that is bewildering; in central London, by contrast, the aerobatic problems facing the huge wheeling flocks of Starlings as they come in to roost in the fading light on a winter evening. Birds can usefully be watched, counted and compared in the different types of countryside passed through on a long train journey (which is really a form of *transect*); or every day on a short commuting trip (which is a form of *census*). You are more likely to see Kestrels on a car journey by motorway than elsewhere (because the long grass of the verges is ideal for hunting voles), though sadly you cannot take the opportunity to study them closely while they hover!

Implicit in this is the idea that birdwatching should never be a case of "Oh, there was only a Blue Tit". Naturally, individual birdwatchers will differ in their approach, some using binoculars the better to appreciate the Kingfisher-nature of the Blue Tit's crown, others to see just where it is feeding, or how it tackles a peanut. Never, never make the quite common mistake that assumes that, once you take an interest in bird biology or numbers, you are mysteriously prohibited from appreciating and enjoying the beauty and charm of these creatures. On the contrary, the more you find out, the more you enquire, the greater the fascination. Birdwatching can be, if you wish, a completely solo pursuit. Many books are available to guide you if you prefer to teach yourself. Equally, perhaps even better, it can be a co-operative enterprise. Whilst one or two is the ideal number if the watcher wants really good views, properly organised group outings can be immensely helpful to the beginner, not only by providing expert help with field identification, but also in showing you the best local birding spots. So, if you are starting out to investigate birdwatching, first find your local society. These are now very numerous—far too numerous to list. Almost all counties have a County Society, while most large towns either have their own club or an RSPB "members' group", and this is true of a good many villages. Often your local library will be able to give you useful addresses, but in cases of difficulty *The Council for Nature* (c/o Zoological Gardens, Regent's Park, London, NW1) which serves as the co-ordinating body for Natural History Societies, should be able to help.

Clubs and societies like these often run a series of field meetings during the year to places of interest, sometimes a day's coach trip away, and during the winter have a series of indoor meetings with a wide variety of films and speakers—there should be enough to cater for most tastes. If your birdwatching leanings are a little more academic, many *Workers' Educational Associations* (in the Telephone Directory) organise various courses, some for beginners, some for more advanced students, at many centres, one of which is likely to be quite near your home. As with all hobbies, practice helps immensely. It is really surprising how many birdwatchers fail to get the best out of their binoculars for this reason, and are slow at picking up and focusing on a flying bird, or one jumping about in a bush, merely because this doesn't come as second nature—but, given practice, it will.

All the bird books in the world cannot teach you to identify birds quickly if you have no field experience. With birds, there are many little things, difficult to put down in words (and often different for each of us) that make a bird *obviously* what it is—"jizz" again. It may be a springy, buoyant way of flying, or an oddly upright perching position—but once, through field experience, you are familiar with it, it "clicks" instantly in your mind. If you want a real demonstration, try now to decide how you *know* that a female House Sparrow is just that, or jot down in detail the plumage characteristics of a Meadow Pipit.

Clearly you should never be without notebook and pencil, and if you can sketch roughly what you see, so much the better. In the early days, take your binoculars, your field identification guide, and your notebook to as many different habitats as possible. Get to know the regular birds that you can expect at different times of year, and better still, get to know their habits. Which species are confiding, like Treecreepers, and allow a close approach? Which are just the opposite, like Redshanks or Blackbirds, and go shrieking off, alarming everything else before you have had a proper look around? All these facets, put together, help you to achieve a closer contact, and thus more enjoyment, with birds.

You will soon find that low tide on the coast, or in an estuary, is no time for good views of ducks or waders—where, then, are they forced to concentrate when the tide comes in? Follow the flightlines and see for yourself. In wetland areas, particularly, when is the sun going to be shining in your eyes? This is a time best avoided, of course, unless you fancy identifying silhouettes!

Remember courtesy while you do all this. There may be other birdwatchers around (even if you haven't noticed them), so, having seen as much as you want, don't just stand up and walk off—leave as inconspicuously and as quietly as you came. Even more important, *always put the birds first*. Never harass tired migrants or unusual birds for just that bit of a better view, and never take risks with nests, even if they are of very common species. If you are a photographer, the same rules apply—perhaps even more strictly at the nest. If you cannot take a photograph without disturbing the vegetation, *don't take it*! Remember, and observe, the Protection of Birds Acts.

This, really, is part and parcel of ensuring that there will be birds around for your grandchildren to enjoy. Conservation is this, and more. Even within your garden, by putting out food and especially water, by planting trees and shrubs (nothing has a lower bird density than plain, mown, grass) and by putting up nestboxes, you can do your bit.

As taxpayers, we also do a bit towards conservation—but a very small bit it is. Our Government, through the *Nature Conservancy Council*, does purchase and maintain some reserves, but the money budgeted for this task is pitifully small and can only be described as woefully inadequate. It is strange that our elected representatives should feel this way about our environment, for the voluntary, or charitable, bodies that own and maintain nature reserves have wide public support. Which is just as well, because they must shoulder the bulk of the burden. A thousand-acre nature reserve costs very much less than a single jet fighter plane!

Three major voluntary bodies share this duty. *The National Trust* (42 Queen Anne's Gate, London, SW1) has, as well as its better-known stately homes, large areas of our countryside and coast under its protection, some as nature reserves. *The Wildfowl Trust* (Slimbridge, Gloucestershire) has a number of wildfowl refuges in various parts of Britain and, incidentally, keeps collections of wildfowl from all over the world (these, and the wild birds, are well worth seeing for their own sake and in order to become familiar with this group). *The Royal Society for the Protection of Birds* (The Lodge, Sandy, Bedfordshire) also maintains and wardens reserves in all types of habitat in Britain and Northern Ireland. Besides acting as the birdwatchers' "watchdog" in protecting rare breeding birds, such as Ospreys and Avocets, and in helping to ensure that the Bird Protection Acts are enforced (particularly where the taking of eggs or wild birds for caging is concerned), the RSPB is our most powerful lobby, being by far the biggest of the birdwatchers' societies. Members have privileges in visiting reserves, and the Society has a range of film, educational and other services available, as well as a strong and well-organised junior wing, the Young Ornithologists' Club.

On a more local level, most counties now have a County Naturalists' Trust, which owns and maintains a variety of reserves, some for plants especially, some for birds, and so on. Your library should be able to tell you the address of the County Secretary, but if not, the County Trusts' parent body, *The Society for the Promotion of Nature Reserves* (Alford Manor House, Alford, Lincolnshire) certainly will.

With pressures increasing at the present rate, not only on our birds, but on our whole environment, it is now possible to argue that membership of bodies like the RSPB and your County Trust has become a duty that none of us should shirk—after all, we share that environment with the wildlife.

There remains one other way in which you can help—to join in and share the enjoyments of co-operative studies, and in one sense this is what all the editions of this book have encouraged birdwatchers to do. The accumulation of biological facts about birds can be completely fascinating and absorbing—and we have seen that it should *add* to your enjoyment of birdwatching. Just as nature reserves are so necessary today—areas where "progress" can be realistically and sensibly managed—so too are facts and figures concerning our birds. Without them, how can we arrive at sensible environmental management plans, how can we argue for nationwide (or worldwide) conservation measures, like the control of pollution, restraint in pesticide usage, rational usage of natural resources like land? While this edition is being written, the Western World is in the throes of an energy crisis the like of which we have not seen before. As I write, it seems most probable that a temporary solution will be sought in extracting more fossil fuels from technically difficult and environmentally dangerous areas (out of sight does not imply safety), like the North Sea or even the western Atlantic, and this as fast as possible. Re-cycling of energy or raw materials barely receives a mention. Housing and industrial development show much the same gloomy picture—everywhere, even in remotest Scotland, piecemeal development occurs more or less unchecked. No thought is given to an overall, national strategy which is so desperately needed. Of course man must continue to live in comfort and to be fed, but ecologists are concerned at the ultimate cost if we neglect plans for the future in our anxiety to provide instant remedies for the immediate problems.

This may seem very sombre, and a situation far removed from any assistance that a birdwatcher can give—but this need not be so. We are discussing the quality of our life, our environment, and as birdwatchers the natural aspects of that environment are even more important. Who better than ourselves to defend this quality, or improve it? Certainly it cannot improve if we shrug our shoulders and turn away—and why should we, anyway, if we enjoy our birdwatching. We can co-operate with others without in any way spoiling our enjoyment—on the contrary.

As an individual, you may make lists of the birds you see, carry out regular counts of a certain area, feed the birds in your garden, put up nestboxes and see how many young are reared and how they fare. *The British Trust for Ornithology* (Beech Grove, Tring, Hertfordshire) was founded by birdwatchers who foresaw the value, practicability and enjoyment-potential of co-operation in field studies, and any active birdwatcher should encourage his local club to participate, and, of course, should involve himself. At any one time, the BTO has a number of projects running: some are organised from headquarters, others by individuals or groups of members. Standards vary, both in level and in the time that needs to be devoted to them. For example, the *Garden*

Birds Feeding Survey seeks information from anyone who feeds birds—housewife, housebound or harassed husband. At the moment, these birdwatchers are witnessing the establishment of the Siskin and Reed Bunting as garden birds. Elsewhere we have talked in some detail of the *Common Birds Census* and *Ringing Scheme*—both long-standing permanent BTO enquiries. *The Birds of Estuaries Survey* seeks facts and figures about the vast international hordes of waders and wildfowl that winter on our estuaries or use them as vital staging posts on migration—and these estuaries are just about our most seriously pressed habitat at the present time, especially threatened by land reclamation and water storage barrages. *The Atlas of Breeding Birds project,* completed in 1972 and soon to be published, has been succeeded by the *Habitat Register,* which entails listing and documenting (Domesday Book style) our bird habitats, site by site, so that planners may be better informed *before* they start to make their plans. The *Atlas* is bound to reveal individual species that need investigation—recently the BTO has been concerned with the Twite, Water Pipit and Cirl Bunting, and there are plans to census Britain's rookeries. Lastly, the *Nest Records Scheme* encourages members (stressing the care that is needed) to visit nests and to record the sequence of events there, the eggs, the young, and how many successful fledglings there are. From these figures, we can see how our birds are doing at this critical recruiting stage of their life cycle.

Not only is it pleasant to participate in this way (and often exciting, too, when you hold your breath, in close proximity to a flock of geese, for example, or to avoid disturbing a cock Bullfinch as it approaches along a hedge, or even just before the garden Robin hops onto your hand to take a worm), it is rewarding and immensely satisfying. I can do no better than quote, and agree with, James Fisher's last paragraph:

"Some people like to make paper plans and push them through regardless of distraction. Others like to wander into a subject and let it take them where it will. I hope I have been able to give the paper-plan people some preliminary headings and I hope that for the others I have given the stream of ornithological thought enough momentum for it to dislodge them from their rest. Birdwatchers are in some ways like golfers, who want to convert all their friends to golf. I don't excuse myself for belonging to this class. If this book succeeds in making new birdwatchers, I hope they will have a very happy time, though I don't see how they can help it."

Index

The birds included in the index are those which are dealt with at some length in the book or are the subject of maps, graphs, etc—brief or passing references to species in the text have not been indexed.

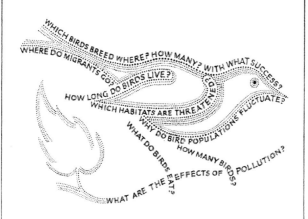

WHICH BIRDS BREED WHERE? HOW MANY? WITH WHAT SUCCESS?
WHERE DO MIGRANTS GO? HOW LONG DO BIRDS LIVE? WHICH HABITATS ARE THREATENED? HOW DO BIRD POPULATIONS FLUCTUATE? WHY DO BIRD POPULATIONS FLUCTUATE? HOW MANY BIRDS? WHAT DO BIRDS EAT? WHAT ARE THE EFFECTS OF POLLUTION?

THE BRITISH TRUST FOR ORNITHOLOGY: a network of thousands of amateur birdwatchers collaborating with a small professional staff to find the facts on which the welfare of our birds depends. The BTO organises field work, runs the Ringing Scheme, maps bird distribution. New investigations include a national habitat register. Whenever birds are threatened the BTO comes forward with proposals and counter-measures based on sound scientific knowledge. All bird lovers are welcome. Active ones for the field work, others for their important moral and financial support. Members receive a quarterly journal *Bird Study* and *BTO News,* an informal bulletin for birdwatchers, seven times a year as well as other privileges.

Write for details of membership to:

BRITISH TRUST FOR ORNITHOLOGY

BEECH GROVE, TRING, HERTS.